Chemical Occurrence Data Sets for Source Water Assessments

The mission of the Awwa Research Foundation (AwwaRF) is to advance the science of water to improve the quality of life. Funded primarily through annual subscription payments from over 1,000 utilities, consulting firms, and manufacturers in North America and abroad, AwwaRF sponsors research on all aspects of drinking water, including supply and resources, treatment, monitoring and analysis, distribution, management, and health effects.

From its headquarters in Denver, Colorado, the AwwaRF staff directs and supports the efforts of over 500 volunteers, who are the heart of the research program. These volunteers, serving on various boards and committees, use their expertise to select and monitor research studies to benefit the entire drinking water community.

Research findings are disseminated through a number of technology transfer activities, including research reports, conferences, videotape summaries, and periodicals.

Chemical Occurrence Data Sets for Source Water Assessments

Prepared by:
Kristyn B. Stevens, **Jose A.H. Sobrinho**, and **Jeffrey S. Rosen**
Technology Planning and Management Corporation
Mill Wharf Plaza, Suite 208, Scituate, Massachusetts 02066

and

Christopher Crockett
Philadelphia Water Department, Philadelphia, Pennsylvania 19107-2994

Sponsored by:
Awwa Research Foundation
6666 West Quincy Avenue, Denver, CO 80235-3098

Published by:

Awwa Research Foundation
Advancing the Science of Water®

IWA Publishing

DISCLAIMER

This study was funded by the Awwa Research Foundation (AwwaRF). AwwaRF assumes no responsibility for the content of the research study reported in this publication or for the opinions or statements of fact expressed in the report. The mention of trade names for commercial products does not represent or imply the approval or endorsement of AwwaRF. This report is presented solely for informational purposes.

Published by IWA Publishing, Alliance House, 12 Caxton Street, London SW1H 0QS, UK

Telephone: +44 (0) 20 7654 5500; Fax: +44 (0) 20 7654 5555; Email: publications@iwap.co.uk;
Web: **www.iwapublishing.com**

AwwaRF report number 90956F. Originally published by AwwaRF for its subscribers in 2003.
IWA Publishing version published 2004.

ISBN 1 84339 865 6

CONTENTS

TABLES

FIGURES

FOREWORD

The Awwa Research Foundation (AwwaRF) is a nonprofit corporation dedicated to the implementation of a research effort to help utilities respond to regulatory requirements and traditional high-priority concerns of the industry. The research agenda is developed through a process of consultation with subscribers and drinking water professionals. Under the umbrella of a Strategic Research Plan, the Research Advisory Council prioritizes the suggested projects based upon current and future needs, applicability, and past work, and forwards the recommendations to the Board of Trustees for final selection. The foundation also sponsors research projects through the unsolicited proposal process; the Collaborative Research, Research Applications, and Tailored Collaboration programs; and various joint research efforts with organizations such as the United States Environmental Protection Agency (USEPA), the U.S. Bureau of Reclamation, and the Association of California Water Agencies.

This publication is a result of one of these sponsored studies, and it is hoped that its findings will be applied in communities throughout the world. The following report serves not only as a means of communicating the results of the water industry's centralized research program but also as a tool to enlist the further support of the nonmember utilities and individuals.

The Foundation's staff and large cadre of volunteers, who willingly contribute their time and expertise, manage projects closely from their inception to the final report. The foundation serves a planning and management function and awards contracts to other institutions such as water utilities, universities, and engineering firms. The funding for this research effort comes primarily from the Subscription Program, through which water utilities subscribe to the research program and make an annual payment proportionate to the volume of water they deliver. Consultants and manufacturers subscribe based on their annual billings. The program offers a cost-effective and fair method for funding research in the public interest.

The Foundation's research agenda addresses a broad spectrum of water supply issues: resources, treatment and operations, distribution and storage, water quality and analysis, toxicology, economics, and management. The ultimate purpose of the coordinated effort is to assist water suppliers to provide the highest possible quality of water economically and reliably. The true benefits are realized when the results are implemented at the utility level. The Foundation's trustees are pleased to offer this publication as a contribution toward that end.

Edmund G. Archuleta, P.E.
Chair, Board of Trustees
Awwa Research Foundation

James F. Manwaring, P.E.
Executive Director
Awwa Research Foundation

ACKNOWLEDGMENTS

The authors of this report are indebted to the individuals within the numerous state departments for their cooperation and participation in this project. We are thankful to the Philadelphia Water Department specially Christopher Crockett for their commitment and involvement in this project. In addition, the authors greatly appreciate the assistance and advice of the AwwaRF Project Manager, Traci Case, and the Project Advisory Committee (PAC) whose members include Ronald Hunsinger, East Bay Municipal Utility District (EBMUD), Glenn Patterson, United States Geological Survey (USGS), and Richard Gullick, American Water Works Service Company (AWWSC). This project benefited from the series of PAC reviews including review of all documents and several website critiques. We are grateful to the PAC for their continued commitment and involvement in this project.

Technology Planning and Management Corporation thanks AwwaRF for its financial and administrative assistance in funding and managing this project.

EXECUTIVE SUMMARY

BACKGROUND

The Safe Drinking Water Act (SDWA), established in 1974, is designed to protect the quality of drinking water in the United States. Congress amended the SDWA in 1996 to address the current source water problems and assist the United States Environmental Protection Agency (USEPA), state agencies, and local water utilities to plan for future drinking water quality challenges and assure the sustainable resource of safe drinking water. Specifically, Section 1453 of the 1996 SDWA amendments requires all states to establish Source Water Assessment Programs (SWAPs). This includes the submission of a plan to the USEPA detailing how the states will delineate source water assessment areas, inventory significant contaminants in these areas, and determine the susceptibility of each public water supply to contamination.

Source water assessments allow states to systematically address issues of potential contamination of public water supplies using existing data from established environmental programs, however the data developed and compiled for source water assessments should be treated as a strategic resource. Each state data set should become part of an overall SWAP inventory, as the data that can be used for SWAPs is located in many different areas and sources and disparate formats. Often times states and local agencies know the data is out there but they may be overwhelmed by how to locate or access it, or simply not have the time to track down the information. The data inventory can serve as a resource to assist agencies in improving regulations and programs, facilitate future research programs, and promote public awareness of the relationship between human activities and the protection of public water supplies.

RESEARCH OBJECTIVES

The collection of existing data on chemical occurrence is one tool to help determine potential contaminants and perform susceptibility analyses for surface and groundwater. As a result, the American Water Works Association Research Foundation (AwwaRF) funded this research project to identify, list, and characterize chemical occurrence databases and data sets that can be used by states and utilities to perform source water assessments. The data sets were inventoried and reviewed based on a variety of descriptors (metadata), including:

- Program title and objectives
- Geographic extent (national, regional, state, watershed, waterbody, or aquifer)
- Time frame
- Chemicals addressed
- Location or description of sampling points
- Sampling methods
- Analytical methods and associated detection limits
- Quality assurance and quality control procedures for sample collection and analysis

- Overall assessment of the quality of the data
- Contact information

APPROACH

This project includes a review of available, relevant source water data sets (e.g., surface water and ground water) that document chemical occurrence in an effort to develop a comprehensive data inventory to serve as a central resource when performing SWAPs. Data sources queried to accomplish this task included national databases and data sets available through regional and state agencies, water utilities and volunteer groups. The majority of databases and data sets included in this research project were identified as a result of the participation and cooperation of local and state agencies. The Project Team also identified national data sets and some state and volunteer groups' data sets through online research.

State agencies filled out surveys at the request of the Project Team and additional data sets were researched in an effort to create a comprehensive data inventory for SWAPs. The questionnaire was designed and standardized to ensure that all state agencies contacted were asked the same questions. The questionnaire (see Appendix C) was posted in the project website http://www.drinkingh2o.com/swap. The website was protected with a password and identification number during the time the research was under development. States participating in the project received authorization and necessary codes to access the project website. This enabled the user to provide source water data set information in an online format. Information compiled through the state survey included data set name and description, program's objectives, funding and sponsoring agencies, period of performance, contact information, and several metadata descriptors (laboratory and sample collection procedures, analytical methods, detection limits and others). Findings were synthesized to reflect data received from each state completing the questionnaire. In addition, national and regional data sets were summarized in the same manner as the state data sets.

The data collected through the state surveys and the online research were stored in a Microsoft database (2000 SQL Server™) and posted in the project website. Through this site, each state participant had the opportunity to test the provided information and evaluate its content and functionality. TPMC is committed to the maintenance of the website for a two-year period and will correct any erroneous entries as well as add any new entries during this time period.

SOURCE WATER QUALITY DATABASE WEBSITE

In an effort to facilitate the SWAP process for states and drinking water utilities, an interactive website was developed to house the source water quality database inventory. The website, http://www.drinkingh2o.com/swap, enables the user to learn more about the national, regional, and state data sets. It also provides the SWAP contact information for each state as well as a direct link to the state SWAP website. The user can also search for data and produce brief and full text data summaries. The website provides some basic cautionary information and recommendations for working with data from disparate data sets. Finally the website allows the user to submit questions, suggestions, corrections, or new data sets through a feedback system developed by the Project Team. This feedback system will be maintained for a period of two years after the public release of the website. The website and the feedback system will be maintained online until July 24, 2004.

UTILITY REVIEW OF THE PROJECT WEBSITE

The Philadelphia Water Department (PWD) and Water Revenue Bureau (WRB) serve the Greater Philadelphia region by providing integrated water, wastewater, and storm water services. The utility's primary mission is to plan, operate, and maintain both the infrastructure and the organization necessary to purvey high quality drinking water, to provide an adequate and reliable water supply for all household, commercial, and community needs, and to sustain and enhance the region's watersheds and quality of life by managing wastewater and storm water effectively.

PWD provided a critique of the website content keeping in mind a few questions. For example, would this website have been useful when they prepared their reports? Is this information valuable? If not, should the focus of the information be changed? PWD also lent their rationale for data mining and the problems and challenges they faced in collecting the data. They have also developed an approach to decide what type of data is appropriate. In addition, PWD documented any lessons learned from their data mining experience. Five major data cautions were recognized during the data inventory review:

- Disparate data formats
- Data duplication
- Data bias and data gaps
- Cautions of the use of the data statistically
- Quality control of the data.

Recommendations were also provided to help the user address these limitations:

- Data mining process
- Estimate of the format of the data and the time to download
- Protocol for database development
- Advice and/or constraints to review the data sets
- Compiling background information on the database

These caveats and recommendations are also posted on the website.

CONCLUSIONS

The product of this research has extensive application to the drinking water and regulatory communities, enriching the ability to assemble and evaluate all existing and readily available water quality related data and information. Specifically, this project documented chemical data stored in a variety of databases, provided a mechanism to access it and the associated metadata. The project results can help organizations that monitor and use chemical water information for the:

- Characterization of source waters and identification of changes or trends in water quality over time
- Identification of existing or emerging water quality problems resulting in increased public health protection against chemical exposures
- Design of specific pollution prevention or remediation programs
- Determination of whether program goals such as compliance with pollution regulations or implementation of effective pollution control actions are being met
- Design monitoring programs

FUTURE RESEARCH

Future efforts related to this information system will include regular updates of both existing databases and new databases as they are located. The project team is committed to maintain and update this information system for a period of two years.

The drinking water community as well as the states overseeing this community should move towards integration of these types of information into readily accessible information systems. This project has taken the first steps towards locating and providing simple interfaces to these data. More robust information systems, which both access and integrate the information could be tremendously valuable to managers and researchers. Appropriate metadata are also critical and should be available in consistent and searchable formats. The metadata will make information generated by a wide range of organizations available for integration and use in management and regulatory development related to safe drinking water.

CHAPTER 1
INTRODUCTION

BACKGROUND

The Safe Drinking Water Act (SDWA) established in 1974 is designed to protect the quality of drinking water in the United States. Congress amended the SDWA in 1996 to assess the current source water problems and assist the United States Environmental Protection Agency (USEPA), state agencies, and local water utilities plan for future drinking water quality challenges and assure the sustainable resource of safe drinking water. Specifically, three main components have been emphasized as the biggest changes:

- Disseminate better information for consumers (e.g., the "right to know");
- Encourage regulatory enhancements, such as improved science, prioritization of effort, and risk assessment; and
- Increased funding for both states and communities through the institution of a Drinking Water State Revolving Fund (SRF).

Specifically, Section 1453 of the 1996 SDWA amendments requires all states to establish Source Water Assessment Programs (SWAPs). This includes the submission of a plan to the USEPA detailing how the states will delineate source water assessment areas, inventory significant contaminants in these areas, and determine the susceptibility of each public water supply to contamination. Currently all 50 states have USEPA approved SWAP plans. Each state is required to have completed source water assessments by 2003.

USEPA has published the *State Source Water Assessment and Protection Programs Final Guidance* (USEPA 1997) to help states develop SWAP submittals. This guidance describes the required content of a SWAP submittal, federal funds available for completion of the assessments, requirements for public participation, and linkages to other federal programs.

Source water assessments allow states to systematically address issues of potential contamination of public water supplies using existing data from established environmental programs, however the data developed and compiled for source water assessments should be treated as a strategic resource. Each state data set should become part of an overall SWAP inventory, as the data that can be used for SWAPs is located in many different areas and sources and disparate formats. Often times states and local agencies know the data is out there but they may be overwhelmed by how to locate or access it, or simply not have the time to track down the information. This project's data inventory can serve as a resource to assist agencies in improving regulations and programs, facilitate future research programs, and promote public awareness of the relationship between human activities and the protection of public water supplies.

OBJECTIVES

The collection of existing data on chemical occurrence is one tool to help determine potential contaminants and perform susceptibility analyses for surface and groundwater. As a result, the American Water Works Association Research Foundation (AwwaRF) funded this research project to identify, list, and characterize chemical occurrence databases and data sets that can be used by states and utilities to perform source water assessments. Data sources queried to accomplish this task included national databases and data sets available through regional and state agencies, water utilities and volunteer groups. The data sets were inventoried and reviewed based on a variety of descriptors (metadata), including:

- Program title and objectives
- Geographic extent (national, regional, state, watershed, waterbody, or aquifer)
- Time frame
- Chemicals addressed
- Location or description of sampling points
- Sampling methods
- Analytical methods and associated detection limits
- Quality assurance and quality control procedures for sample collection and analysis
- Overall assessment of the quality of the data
- Contact information

State agencies filled out surveys at the request of the Project Team and additional data sets were researched in an effort to create a comprehensive data inventory for SWAPs. Finally, this information was synthesized into an interactive website (http://www.drinkinghh2o.com/swap) to serve as a resource for conducting SWAPs.

REPORT ORGANIZATION

This report contains six sections. Chapter 1 presents an introduction, which details the background, statement of the problem, and the project objectives. Chapter 2 presents Methods and Materials, which details the approach taken in this process as well as the relevant data sources. Chapter 3 summarizes the website content and discusses the data inventory results. Chapter 4 presents the Philadelphia Water Department (PWD) review and critique on the website. This chapter also discusses limitations and cautions to be aware of when working with disparate data sets and provides recommendations to address the given limitations. Chapter 5 provides a summary of this project. Chapter 6 discusses the relevance of this project to the drinking water industry. This report also contains 6 appendices: Appendix A, a State contact list; Appendix B, an invitation letter requesting state participation in the project; Appendix C, the state survey used in the data collection process; Appendix D, summary of the national databases; Appendix E, summary of the regional data sets; and Appendix F, summary of the states data sets.

CHAPTER 2
METHODS AND MATERIALS

APPROACH

This project includes a review of relevant source water data sets (e.g., surface water and ground water) that document chemical occurrence, in an effort to develop a comprehensive data inventory to serve as a central resource when performing SWAPs. The majority of databases and data sets included in this research project were identified as a result of the participation and cooperation of local and state agencies. The Project Team also identified national data sets and some state and volunteer groups' data sets through online research. For the state data sources, key personnel in charge of source water assessments, source water protection or drinking water programs were identified and contacted in all fifty states. This initial interaction with the states was critical to select the most appropriate point of contact to help the Project Team locate, evaluate, and obtain source water data. A state contact list is provided in Appendix A for all 50 states. Following this step, a letter was sent to the state contact requesting their participation in the project. An example of the state letter is included in Appendix B.

A questionnaire was designed and standardized to ensure that all agencies contacted were asked the same questions. The questionnaire (see Appendix C) was posted in the project website http://www.drinkingh2o.com/swap. The website was completely protected with a password and identification number during the time the research was under development. States participating in the project received authorization and necessary codes to access the project website. This enabled the user to provide source water data set information in an online format.

Information compiled through the state survey included data set name and description, program's objectives, funding and sponsoring agencies, period of performance, contact information, and several metadata descriptors (laboratory and sample collection procedures, analytical methods, detection limits and others). Findings were synthesized to reflect data received from each state completing the questionnaire. In addition, national and regional data sets were summarized in the same manner as the state data sets.

In addition to the data provided by each agency the Project Team researched additional data sets. For those states whose input to this project was minimal, the Project Team reviewed the various state agency websites and used Internet search engines to try and develop a comprehensive data inventory. The results of this website is a combination of the two approaches.

The data collected through the state surveys and the online research were stored in a Microsoft database (2000 SQL Server™) and posted on the project website. Through this site, each state participant had the opportunity to test the provided information and evaluate its content and functionality. TPMC is committed to the maintenance of the website for a two-year period and will correct any erroneous entries as well as add any new entries during this time period.

Quality assurance and quality control (QA/QC) procedures were performed throughout the evaluation of the different data sources in an effort to identify the most reliable and complete data sets. Each data set was carefully evaluated for metadata completeness as well as internal QA/QC procedures.

DATA SOURCES

There are so many sources of water quality data for surface and groundwater drinking water sources. Source water data sets store monitoring data developed and implemented by state, local, and federal agencies and private entities to assess and evaluate the quality of surface and groundwater. In addition, many other sources of information can also contribute to SWAPs. For instance, Section 303(d) of the Clean Water Act (CWA) requires states to identify and list impaired or threatened waters. By requirement, these lists are submitted to USEPA for review and approval every April of even years. Once the listing process begins, it initiates Total Maximum Daily Loads (TMDL) requirements. TMDLs assist with the regulation of both point and non-point source discharges in an impaired waterbody. Datatsets used to create TMDLs are also of use to SWAPs. Monitoring programs related to the combined sewer overflow management programs, point source programs like National Pollutant Discharge Elimination System (NPDES), and non-point source programs such as Better Assessment Science Integrating Point and Nonpoint Sources (BASINS) are also of interest.

The following section describes and summarizes existing data sets that contain potentially useful information about chemical occurrence in source waters. Summaries for all databases and data sets identified through this study are also included in the CD-ROM to be distributed with this report. Depending on the geographic extent covered by the monitoring programs, the data set may be classified as national, regional or state.

National Databases

Chemical data are available in several national databases, which are complex systems used to store and manage various data sets. The national databases identified through this research include Storage and Retrieval (STORET), Safe Drinking Water Information System (SDWIS), National Water Information System (NWIS), National Water Quality Assessment (NAWQA) Program, Nutrient Criteria, National Contaminant Occurrence Database (NCOD) and National Stream Quality Accounting Network (NASQAN). Appendix D summarizes the national databases providing a short data set description, active Uniform Resource Locator (URL), chemical groups, period of performance and contact information for each data set.

Storage and Retrieval (STORET)

STORET (http://www.epa.gov/storet), including the STORET Legacy Data Center (LDC) and the Modernized STORET, is the USEPA's largest computerized environmental data system. Both systems, LDC and STORET, are online databases that contain parametric water quality, biological, and physical monitoring data collected within the fifty states, territories, and jurisdictions of the U.S., along with portions of Canada and Mexico. The LDC contains historical water quality data dating back to the early part of the 20th century and collected up to the end of 1998. Modernized STORET, or simply STORET, stores data collected in the beginning of 1999, along with older data that has been properly documented and migrated from the LCD. This data is typically generated by state, local, and federal agencies and private entities. STORET contains metadata including sample location (e.g., latitude, longitude, state, county, Hydrologic Unit Code (HUC) and a brief site identification), when the sample was collected, the medium sampled (e.g., water, sediment, fish tissue), and the sponsoring organization. STORET also contains information about why the data was collected, sampling methods, and analytical methods (including the laboratory conducting the analyses). It also stores information related to the quality control checks used when sampling, handling the samples, and analyzing the data; and the personnel responsible for the data.

Safe Drinking Water Information System (SDWIS)

SDWIS (http://www.drinkingh2o.com/swap/html/national.htm#sdwis) is the USEPA's national regulatory compliance database storing information on the Nation's approximately 175,000 public water systems and violations of drinking water regulations. Designed to replace the system known as FRDS (Federal Reporting Data System), SDWIS contains information reported by states as required by the SDWA. The system allows the user to locate the drinking water supplier of interest and review any violations and relevant enforcement history over the past decade. The user can then locate systems either by entering the water identification system number or by selecting the geographic area of interest and defining the water system name, county, and/or population served.

National Water Information System (NWIS)

NWIS (http://www.drinkingh2o.com/swap/html/national.htm#nwis) is an online national database (created by the United States Geological Survey (USGS) to serve as a repository for the nation's water quality data. NWIS is a compilation of several other USGS databases, which includes information on sampling sites, groundwater levels in wells, and water flow and levels in streams, lakes, and springs. Also stored in the database is the chemical and physical data for streams, lakes, springs, and wells. Some of the data from selected surface and ground waters are transmitted and available online in real-time.

National Water Quality Assessment (NAWQA) Program

The USGS NAWQA Program (http://water.usgs.gov/nawqa/) provides data about water chemistry, hydrology, land use, stream habitat, and aquatic life in more than 50 major river basins and aquifers covering nearly all fifty states. The program collects data on nutrients, pesticides, volatile organic compounds, and trace elements, providing a framework for national and regional water quality assessments. These assessments will identify water quality changes and trends and their causes. The first two years of any NAWQA Program involve compilation and analysis of existing data on water quality conditions, hydrology, geology, and changes in land-use activities and management practices. In addition to USGS data, information and methods developed by other federal agencies, as well as by state and local agencies, universities, and volunteer organizations, are reviewed and integrated as appropriate. A subsequent three-year period of data collection and analysis follows the first step of collection of preliminary information. For each study unit, 3- to 5-year periods of intensive data collection and analysis will be alternated with 5- to 6-year periods of less intensive study and monitoring.

National Contaminant Occurrence Database (NCOD)

The NCOD (http://www.drinkingh2o.com/swap/html/national.htm#ncod) supports the USEPA in identifying contaminants for regulation and subsequent regulation development. The NCOD contains occurrence data from both Public Water System Supervision (PWSS) Program and other sources (like the USGS NWIS) on physical, chemical, microbial and radiological contaminants. Query results include summaries but no actual analysis; users can build a set of data to download and analyze for a specific purpose. The user can request information regarding the quality of the local drinking water, any violations reported by the local water system of interest, and what types of contaminants can be found within the local system.

National Stream Quality Accounting Network (NASQAN)

NASQAN (http://www.drinkingh2o.com/swap/html/national.htm#nasqan) is a USGS database containing water chemistry and sediment data collected from several major rivers throughout the U.S. Physical and chemical measurements made at NASQAN stations include basic water quality, major ions, nutrients, pesticides, and suspended sediments. Aggregation of comparable information obtained through data analysis from the study units will be used to explain differences and similarities in observed water quality conditions among study areas. Data can be obtained through a text based or a map interface. It will also display active and inactive monitoring data.

Nutrient Criteria

USEPA developed the online Nutrient Criteria database (http://www.drinkingh2o.com/swap/html/national.htm#nc) to store and analyze nutrient water quality data and to serve as resource for states, tribes, and others in establishing scientifically defensible numeric nutrient criteria. It contains ambient data converted from the STORET LDC, USGS NASQAN, and NAWQA. Other sources include universities, states, and tribes. The ultimate purpose of the data is to develop ecoregional waterbody-specific numeric nutrient criteria. This database will also be used by USEPA to develop ecoregionally representative nutrient criteria for all waterbodies of the United States.

Other National Database Tools

There are additional national databases that provide an overall data assessment including contaminant levels and indicators of the water sources and watersheds but they do not provide the raw chemical data. This includes Better Assessment Science Integrating Point and Nonpoint Sources (BASINS), Surf Your Watershed, EMPACT (Environmental Monitoring for Public Access and Community Tracking), Envirofacts, and WATERS (Watershed Assessment, Tracking & Environmental Results). Each database is listed and summarized below.

Better Assessment Science Integrating Point and Nonpoint Sources (BASINS)

BASINS (http://www.epa.gov/ost/basins/) is a computer-based environmental analysis system developed by USEPA for use by regional, state, and local agencies in performing watershed- and water-quality-based studies. It was developed to address three objectives:

- To facilitate examination of environmental information;
- To support analysis of environmental systems; and
- To provide a framework for examining management alternatives.

BASINS integrates a geographic information system (GIS), national watershed data, and state-of-the-art environmental assessment and modeling tools into one package to quickly assess large amounts of point and nonpoint source data. In addition to several products for watershed assessment, the database stores statistical summaries of water quality monitoring for 50 physical and chemical-related parameters. Parameter-specific statistics are computed by each station in 5-year intervals from 1970 to 1994.

Surf Your Watershed

Surf Your Watershed (http://www.epa.gov/surf/) is a database developed by the USEPA that organizes and presents aquatic resource information for approximately 2,000 watersheds in the 50 US states. This database contains several environmental data products developed from existing national water quality databases. The database provides overall assessment (contaminant level and indicators) rather than raw data for water sources and watersheds. The strength of this data set is that it contains documentation of the data sources used and helps the user to locate valuable source data.

EMPACT (Environmental Monitoring for Public Access and Community Tracking)

The EMPACT (http://www.epa.gov/empact/) program was created to bring people current (time-relevant) environmental information in a user-friendly format. USEPA is supporting the EMPACT program by enabling selected urban areas to monitor, interpret, and communicate real-time data. Some of these areas include Des Moines, Washington, D.C., Cincinnati, and New Orleans. Online links to these programs are listed below. Aggregated results from physical and chemical data can be found at most of the EMPACT sites for the selected locations.

- Des Moines Water Works, IA (http://www.dmww.com/empact.asp)
- Chesapeake Bay, Washington, DC (http://mddnr.chesapeakebay.net/empact/)
- Ohio River, OH (http://www.orsanco.org/empact/)
- Jefferson Parish, LA (http://www.jeffparish.net/pages/index.cfm?DocID=1228)

Envirofacts

The Envirofacts Data Warehouse (http://www.epa.gov/enviro/index_java.html) provides the public with direct access to the USEPA's databases. Users may retrieve environmental information from databases pertaining to Air, Chemicals, Facility Information, Grants/Funding, Hazardous Waste, Superfund, Toxic Releases, Water Permits, Drinking Water, Drinking Water Contaminant Occurrence, and Drinking Water Microbial and Disinfection Byproduct Information (Information Collection Rule [ICR]). Online queries allow data to be retrieved from several databases at once, or from one database at a time, and users can create reports or generate maps of environmental information using several Mapping Applications. Information about the spatial data used by Mapping Applications is also available.

WATERS (Watershed Assessment, Tracking & Environmental Results)

WATERS (http://www.epa.gov/waters/) was developed by USEPA to quickly identify and generate summary reports on the status of individual waterbodies of a state. Since WATERS is a compilation of two USEPA databases (Water Quality Standards and TMDL Tracking System) and the USGS National Hydrographic Data set (NHD), the database is able to provide information on all impaired waters listed under section 303(d) of the Clean Water Act, including their designated uses and criteria. After completion, the database will provide information from all 50 states and will be linked to sources containing data about ambient and drinking water quality, polluted runoff, fish consumption advisories, etc.

Regional And State Data Sets

This section identifies and summarizes regional and state data sets, which store information on several chemical parameters useful for source water assessments. The data stored in these systems were collected for a variety of programs, and the data quality, standards used for analytical work, and spatial data coverage may vary widely. As shown in Table 2.1, this research identified 9 regional and 71 state data sets. The regional and state data sets are summarized in the Appendix E and F, respectively, including a brief description, URL, chemical groups, period of performance and contact information. The components of each national, regional and state data set will be evaluated and characterized in the following chapter.

Table 2.1
Regional and state data sets

Geographic Extent	State	Database/Data set name
Regional	Arkansas, Louisiana, Missouri, Oklahoma, and Texas	Arkansas STORET
Regional	Connecticut, Massachusetts, Maine, New Hampshire, Rhode Island, and Vermont	New England Interstate Water Pollution Control Commission (NEIWPCC)
Regional	Delaware, New Jersey, New York, and Pennsylvania	Schuylkill and Delaware River Watershed Database
Regional	Kansas, Missouri, Nebraska, Oklahoma	Kansas Stream Chemistry Database
Regional	Maryland, Pennsylvania, and Virginia	Alliance Citizen Monitoring Program Database
Regional	Maryland and Virginia	Chesapeake Bay Toxics Program Database
Regional	Maryland and Virginia	Chesapeake Bay Program Water Quality Database
Regional	Maryland and Virginia	Occurrence and Distribution of Pesticides in Chesapeake Bay
Regional	Oklahoma, Nebraska, and Nevada	Groundwater Protection Council (GWPC) Environmental Information Management Suite (EIMS)
State	Alabama	Reservoir Monitoring Data
State	Alaska	State of Alaska Public Water System Supervision (PWSS) Database
State	Arizona	Water Quality Division Database
State	California	Department of Pesticides Regulation's Surface Water Monitoring Database
State	California	Department of Water Resources Water Data Library
State	California	Malibu Creek Stream Team Database
State	Colorado	Rivers of Colorado Water Watch Network Database
State	Delaware	Delaware Public Water Supply Systems Database
State	Delaware	Schuylkill and Delaware River Watershed Database
State	Florida	Generalized Water Information System (GWIS3)
State	Georgia	Safe Drinking Water Information System/State
State	Georgia	Radold
State	Hawaii	Department of Health Water Quality
State	Idaho	Idaho Surface Water Quality Statewide Network
State	Illinois	Illinois Source Water Assessment Database
State	Illinois	Radium in Groundwater from Public Supply Aquifers in Northern Illinois
State	Illinois	The Lower Illinois River Basin NAWQA Study Unit
State	Indiana	Drinking Water Compliance Database and SDWIS/State
State	Iowa	Environmental Monitoring for Public Access and Community Tracking/Des Moines Database
State	Iowa	Iowa STORET
State	Kansas	Kansas Groundwater Quality Database
State	Kansas	Kansas Lake and Wetland Water Quality Database

(continued)

Table 2.1 (continued)
Regional and state data sets

Geographic Extent	State	Database/Data set name
State	Kentucky	Compass
State	Kentucky	Ledbetter and Panther Watersheds Database
State	Kentucky	The Kentucky Lake Long-Term Monitoring Program Data set
State	Kentucky	The Lake Barkley Database
State	Maine	Public Educational Access to Environmental Information in Maine (PEARL)
State	Maine	Public Water Supply Water Quality Data
State	Maryland	Drinking Water Monitoring Data - SDWIS
State	Maryland	Maryland Department of the Environment (MDE) / Department of Natural Resources (DNR) Core Trend Data
State	Massachusetts	Site/Reportable Releases Look-Up
State	Massachusetts	Underground Storage Tank (UST) Registry
State	Massachusetts	Water Quality Testing System
State	Michigan	Act 452 Part 201 Environmental Remediation Sites
State	Michigan	Leaking Underground Storage Tanks
State	Michigan	National Pollutant Discharge Elimination System (NPDES) Permit List
State	Michigan	WaterChem Database
State	Minnesota	Groundwater Monitoring and Assessment (GWMAP) Program Database
State	Minnesota	Nitrate Nitrogen Analyses from Private Water Supply Wells
State	Minnesota	Public Water Supply Supervision Program Compliance Monitoring
State	Minnesota	Statewide Surface Water Monitoring for Nutrients and Pesticides
State	Mississippi	Source Water Assessment Program (SWAP) Geographic Information System (GIS)
State	Missouri	Missouri Stream Team Database
State	Montana	Montana Natural Resource Information System
State	Nebraska	Quality-Assessed Agrichemical Contaminant Database for Nebraska Groundwater
State	Nevada	State of Nevada Surface Water Monitoring Network – Lakes and Reservoirs
State	Nevada	State of Nevada Surface Water Monitoring Network - Streams
State	New Hampshire	Watershed Surface Water Quality Data
State	New Jersey	New Jersey United States Geological Survey
State	North Dakota	Water Resources Data – Ground/Surface Water Data
State	Ohio	Miami Conservancy District – Groundwater Quality Data
State	Oklahoma	Oklahoma Environmental Information Management Suite

(continued)

Table 2.1 (continued)
Regional and state data sets

Geographic Extent	State	Database/Data set name
State	Oregon	Laboratory Analytical Storage and Retrieval (LASAR) Database
State	Oregon	Oregon Water Quality Index
State	Oregon	Oregon Safe Water System: Chemical Detections, Burns Water Department
State	Pennsylvania	Pennsylvania Spatial Data Access (PASDA) – The Pennsylvania Geospatial Data Clearinghouse
State	Pennsylvania	Susquehanna River Basin Commission (SRBC) Nutrient Assessment Program
State	Rhode Island	Rhode Island Public Water Inventory
State	South Carolina	South Carolina Ambient Ground Water Quality Monitoring Network
State	South Carolina	South Carolina Groundwater Contamination Inventory
State	Texas	Source Water Assessment Program (SWAP) Potential Sources of Contamination (PSOC) Data sets
State	Texas	Texas Natural Resources Conservation Commission (TNRCC) – Sampling Data Query, Surface Water Quality Monitoring
State	Texas	Texas Water Development Board (TWDB) Groundwater Database (GWDB)
State	Utah	Statewide Geographic Information Database
State	Utah	Underground Storage Tanks (UST), Leaking Underground Storage Tanks (LUST), Brownfields, Superfund, Toxics Release Inventory (TRI), Voluntary Clean Program (VCP)
State	Vermont	Water Supply Division (WSD) Inventory
State	Washington	Environmental Assessment Program – River Stream and Water Quality Monitoring
State	Washington	Pesticides in Public Supply Wells of Washington State
State	West Virginia	West Virginia Source Water Assessment and Wellhead Protection Program SWAP data online
State	Wisconsin	Drinking Water System
State	Wyoming	Statewide Water Resources Data Inventory

CHAPTER 3
SOURCE WATER QUALITY DATABASE WEBSITE

OBJECTIVES

In an effort to facilitate the SWAP process for states and drinking water utilities, an interactive website was developed to house the source water quality database inventory. The website enables the user to learn more about the national, regional, and state data sets. The user can search for data and produce brief and full text data sets summaries. In addition to information on data sets with chemical occurrence, this tool provides some basic cautionary information and recommendations for working with data from disparate data sets. It also provides the SWAP contact information for each state as well as a direct link to the state SWAP website. Finally the website allows the user to submit questions, suggestions, corrections, or new data sets through a feedback system developed by the Project Team. This feedback system will be maintained for a period of two years after the public release of the website. The website and the feedback system will be maintained online until July 24, 2004.

WEBSITE COMPONENTS

The project website consists of ten main pages:

- Home page
- National
- Regional
- State
- Search
- Summaries
- Caveats
- Feedback
- Help
- State Survey

The components for each page are addressed in the following paragraphs. Graphical depictions from the website are included to facilitate the understanding of the website organization. The website can be accessed at http://www.drinkingh2o.com/swap.

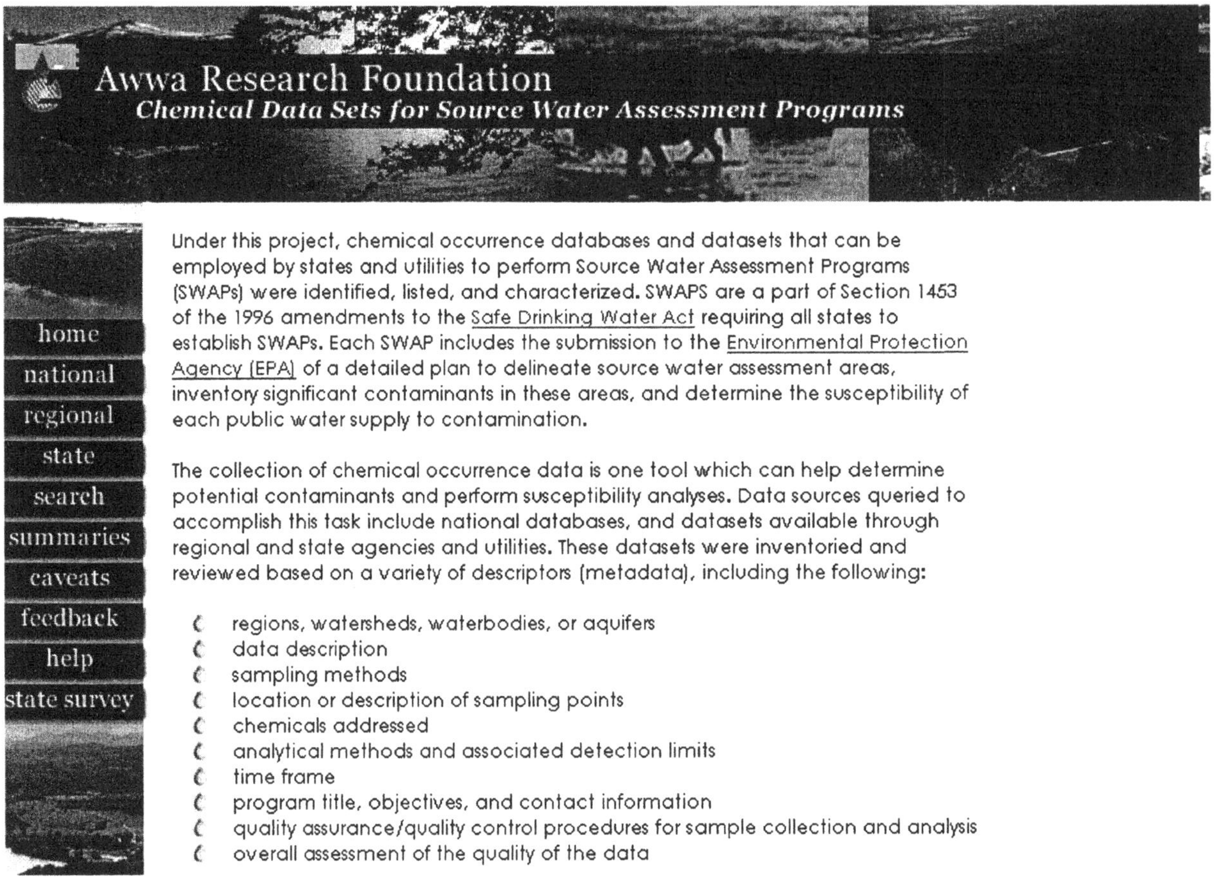

Figure 3.1 Home Page for the project website

The home page (Figure 3.1) provides a brief background of the project as well as the main objectives of the project. It also provides a list of the descriptors (metadata) used to evaluate each data set including program title and objectives, geographic extent, time frame, location, chemical groups, and contact information for the data set. Sampling methods, analytical methods and associated detection limits, quality assurance and quality control procedures for sample collection and analysis, and an overall assessment of the quality of the data are also evaluated.

National

The national page (Figure 3.2) contains summaries and links to the major national databases and several tools for assessing chemical data. At the top of the page is a list of the databases and tools. The user can click on an item to view its summary. In addition, clicking on the title above the summary will transfer the user to that database's or tool's web interface.

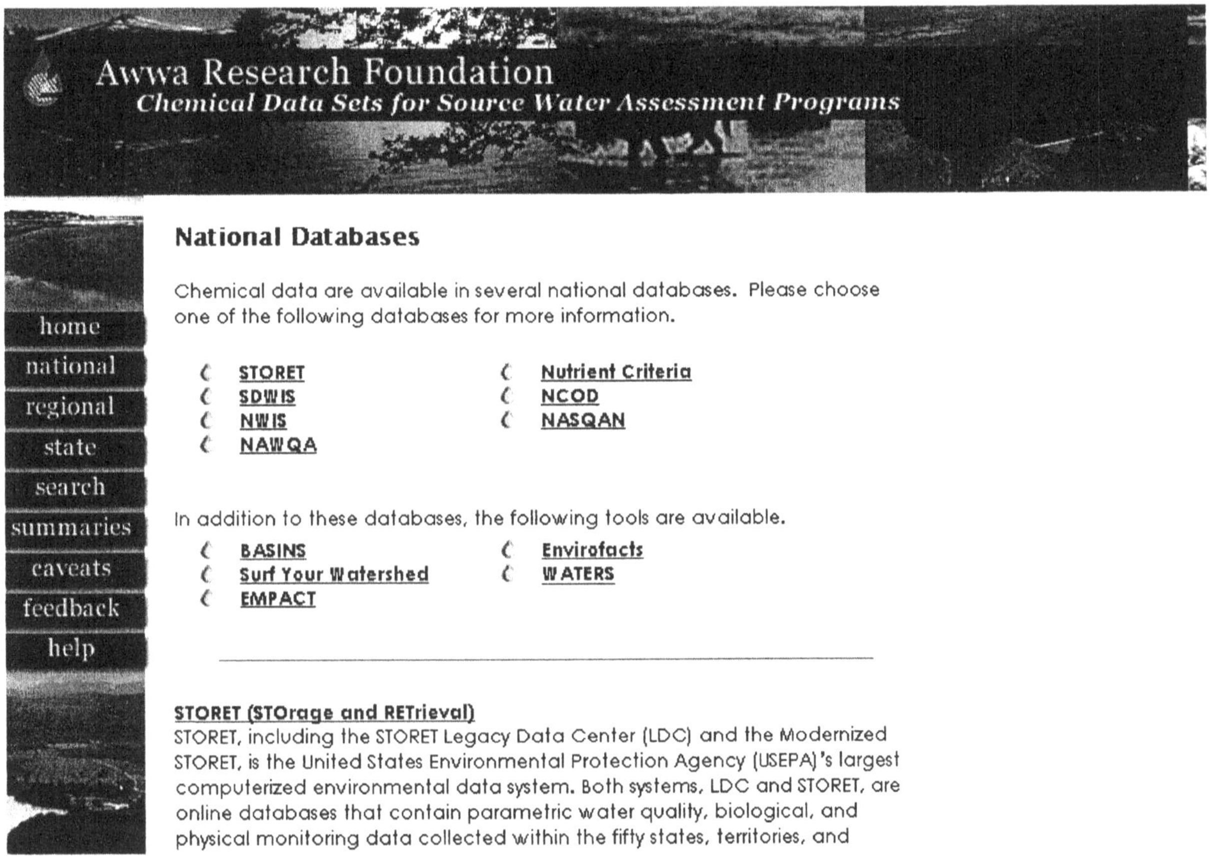

Figure 3.2 National database web page

Regional

The regional page (Figure 3.3) contains summaries for the regional data sets. The regional page only lists a brief data set summary including description, geographic extent, chemical groups, and period of performance. The user can click on the title of any data set to view a full text data set summary.

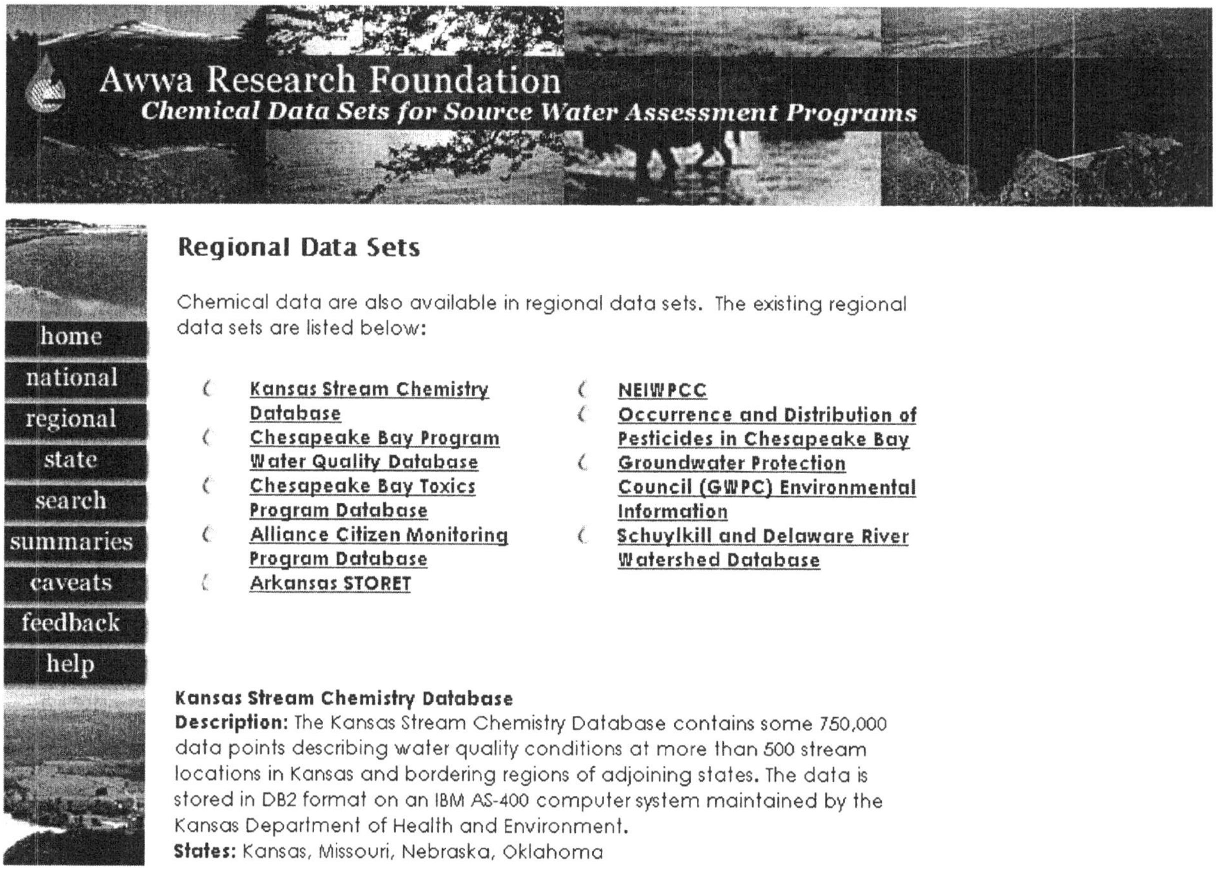

Figure 3.3 Regional data set web page

State

The state page (Figure 3.4) allows the user to view summaries of the data sets available for each state. Clicking on a state (either on the map or using the text links below it) will bring up a brief summary of data sets available for that state, including the program name, description, chemical groups, period of performance, and contact information. A full summary of information on that data set is available by clicking on the data set name.

A link to SWAP resources for the state is also provided at the bottom of each state page. If the state's SWAP program is not available online, a link is provided to the state's environmental agency or a water quality monitoring website. Some links allow the user to download PDF files. To read such files, the user must have the Adobe Acrobat Reader.

Figure 3.4 State data set web page

Search

The Search page (Figure 3.5) allows the user to query all source water information contained in the data inventory. If the users wish to view all records, they may simply click on the "submit" button without making any selections. The following fields are available to search existing records: Geographic Extent, USEPA Region, State, Chemical Group, and Project Start/End Dates. The same information contained in the search mechanism can be selected through the summaries page as well.

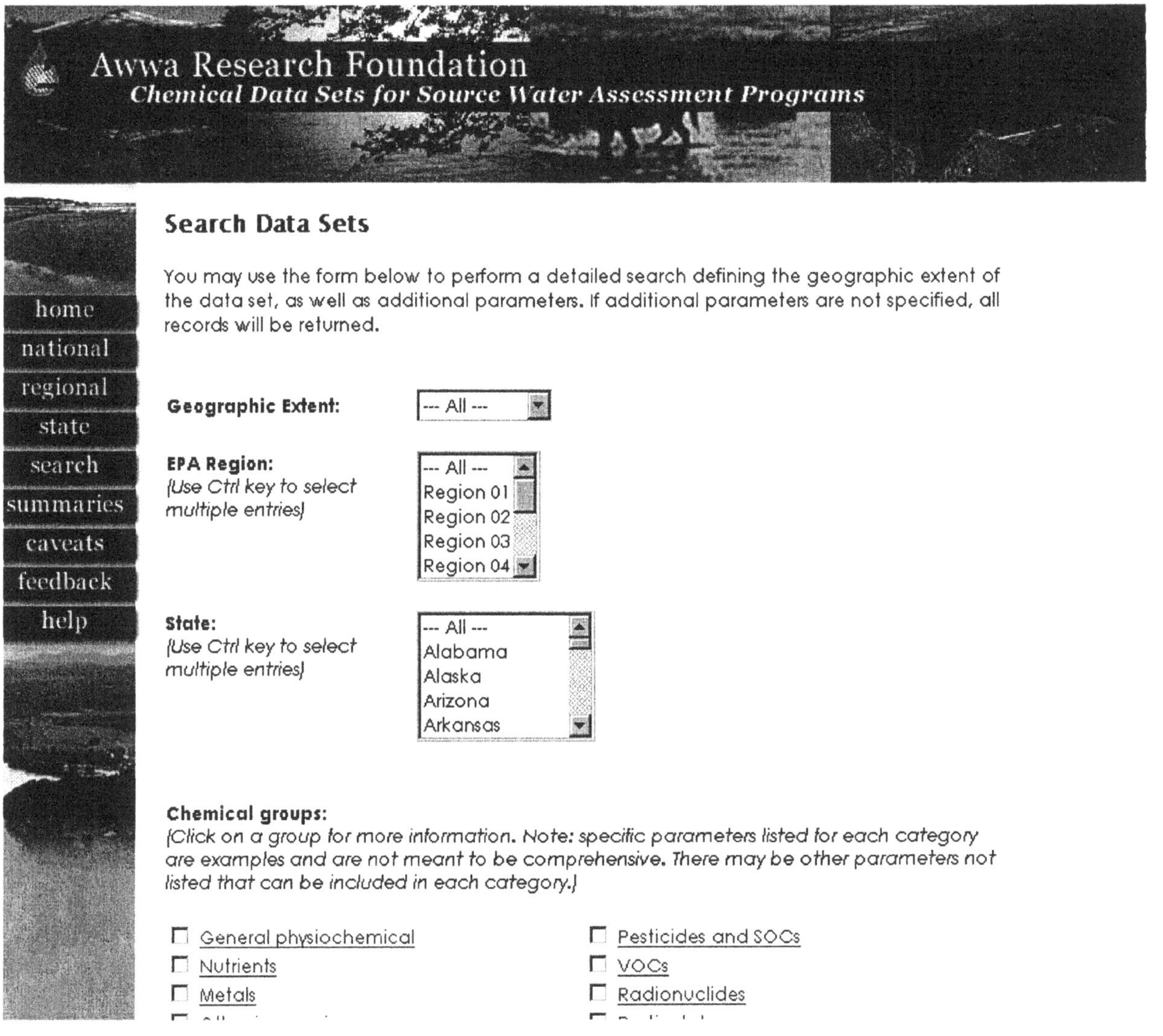

Figure 3.5 Search engine web page

The user can choose from drop-down lists to further limit search results. For instance, the user may choose to utilize the chemical group selections. The checkboxes are used to select as many parameters as desired. To obtain more information on a chemical group, the user can click on the group name and a window will pop up with further detail (Figure 3.6).

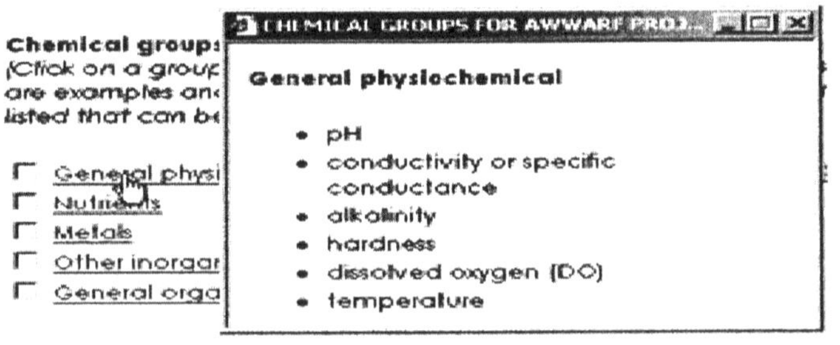

Figure 3.6 Example of chemical groups

In this example the user selected the chemical group *General Physiochemical*. The pop-up window opens with more information on a representative group of parameters that fall under this group: pH, conductivity or specific conductance, alkalinity, hardness, dissolved oxygen (DO), and temperature. In addition to the basic water chemistry group, the chemical groups included nutrients, metals, other inorganics, general organics, pesticides and synthetic organic compounds (SOCs), volatile organic compounds (VOCs), radionuclides and particulates. A list of the chemical groups and representative parameters follows below. The parameters are examples and they are not meant to be comprehensive. There may be other parameters not listed that can be included under each category.

Nutrients:

- Ammonia
- Organic nitrogen
- Nitrite
- Nitrate
- Orthophosphate
- Organic phosphorous
- Polyphosphate

Metals:

- Zinc
- Copper
- Iron
- Manganese
- Nickel
- Chromium
- Cadmium
- Lead
- Mercury

Other inorganics:

- Arsenic
- Asbestos (silicate minerals)
- Sulfate
- Fluoride
- Cyanide
- Perchlorate
- Bromide
- Chloride

General organics:

- Biochemical oxygen demand (BOD)
- Chemical oxygen demand (COD)
- Total organic carbon (TOC)
- Dissolved organic carbon (DOC)
- Disinfection by-product (DBP) precursors
- Algal toxins

Pesticides and SOCs (and their degradation products):

- Atrazine
- Endrin
- Aldrin
- Lindane
- Chlordane
- 2, 4-D
- Polychlorinated biphenyls (PCBs)
- Dioxins
- Pharmaceuticals

VOCs (and their degradation products):

- Benzene
- Toluene
- Ethyl benzene
- Xylenes
- Carbon tetrachloride
- Tetrachloroethylene (PCE)
- Trichloroethylene (TCE)
- Vinyl chloride
- Trichlorobenzene
- 1,1,1-trichloroethane

- Dichloromethane
- Methyl tert-butyl ether (MTBE)
- Trihalomethanes (THMs) (i.e., chloroform, bromodichloromethane, dibromochloromethane, bromoform)

Radionuclides:

- Radon
- Radium
- Uranium
- Alpha activity
- Beta activity

Particulates:

- Suspended solids
- Total solids
- Dissolved solids
- Particle counts
- Turbidity

The user can also select project start/end dates: From the drop-down lists on the left of the screen, the user may choose Before, On, or After, and enter the dates they would like to search in the field on the right. In the example below, the user has chosen to view all records occurring before January 1, 2000 (Figure 3.7).

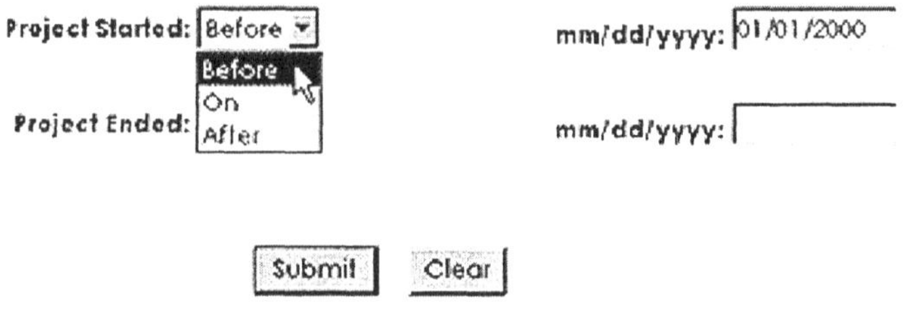

Figure 3.7 Example of project start/end dates

Once the user has completed entering information, click on the "submit" button at the bottom of the screen to begin the search, or click on the "clear" button to start again. After clicking on the "submit" button, all results meeting the selected criteria will appear on the "Search Results" page. The names of matching databases are displayed along with the program name. A full summary of information is available by clicking on the database name in the first column (Figure 3.8).

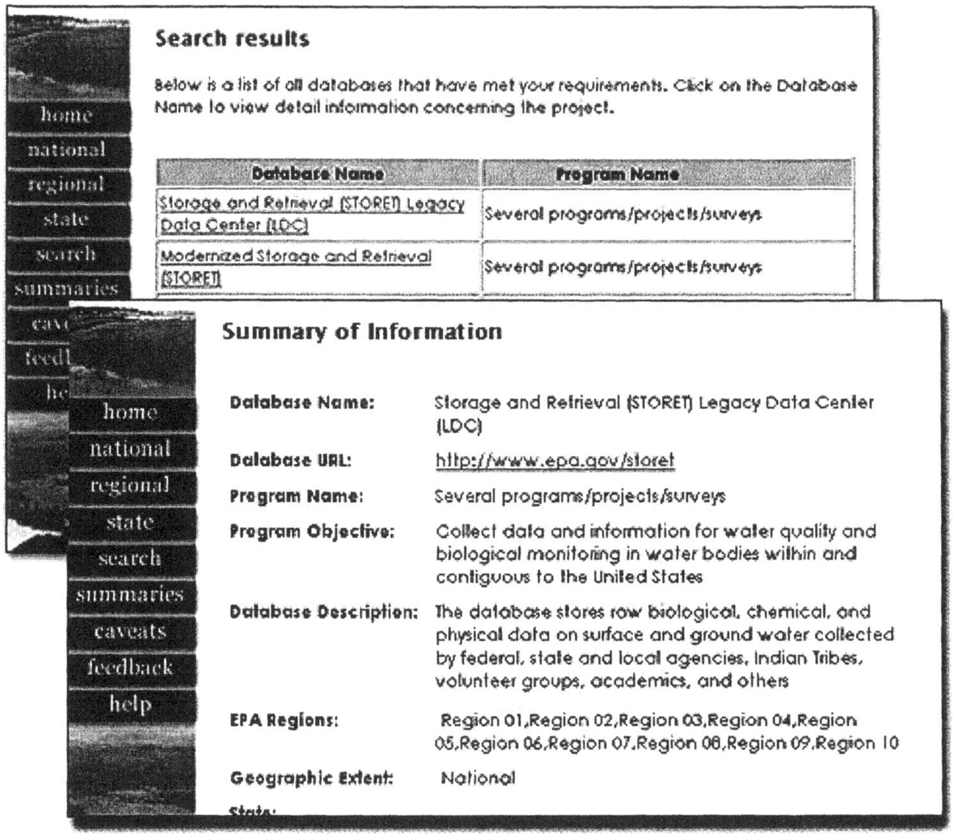

Figure 3.8 Search results web page

If the initial search does not return any results, the "back" button can be used to return to the search page. Be sure that all keywords have been spelled correctly, and that all dates are in the mm/dd/yyyy format. If these are correct, the user may modify the additional search parameters before clicking on the "submit" button to search again.

Summaries

Through this page, the user can select a state from the drop-down list to display a full summary of information for that state. Alternatively "national" can be selected to display a summary of information for all national data sets (Figure 3.9). This is the same full text information that the search engine produces.

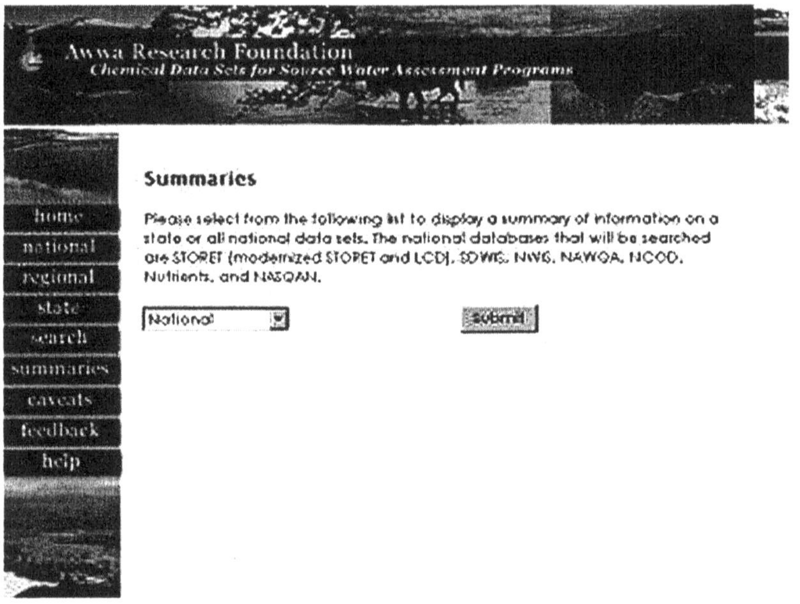

Figure 3.9 Summaries web page

Caveats

This page (Figure 3.10) provides the limitations identified in the utility website review and recommendations for addressing the given limitations. This is critical as data duplication and data compatibility can bias any statistical investigation and must be considered when using data from disparate data sources.

Feedback

"Feedback" allows the user to send email to the developers of the source water quality database website. From this page the user can click on the email link to create a new email message using their default mail program. Simply fill in the body of the email and send to submit comments, questions, or suggestions about the program or this website. If the user is aware of additional data sets not included on this website or necessary data corrections to existing data sets, the user should click on the "Feedback" button to report this as well.

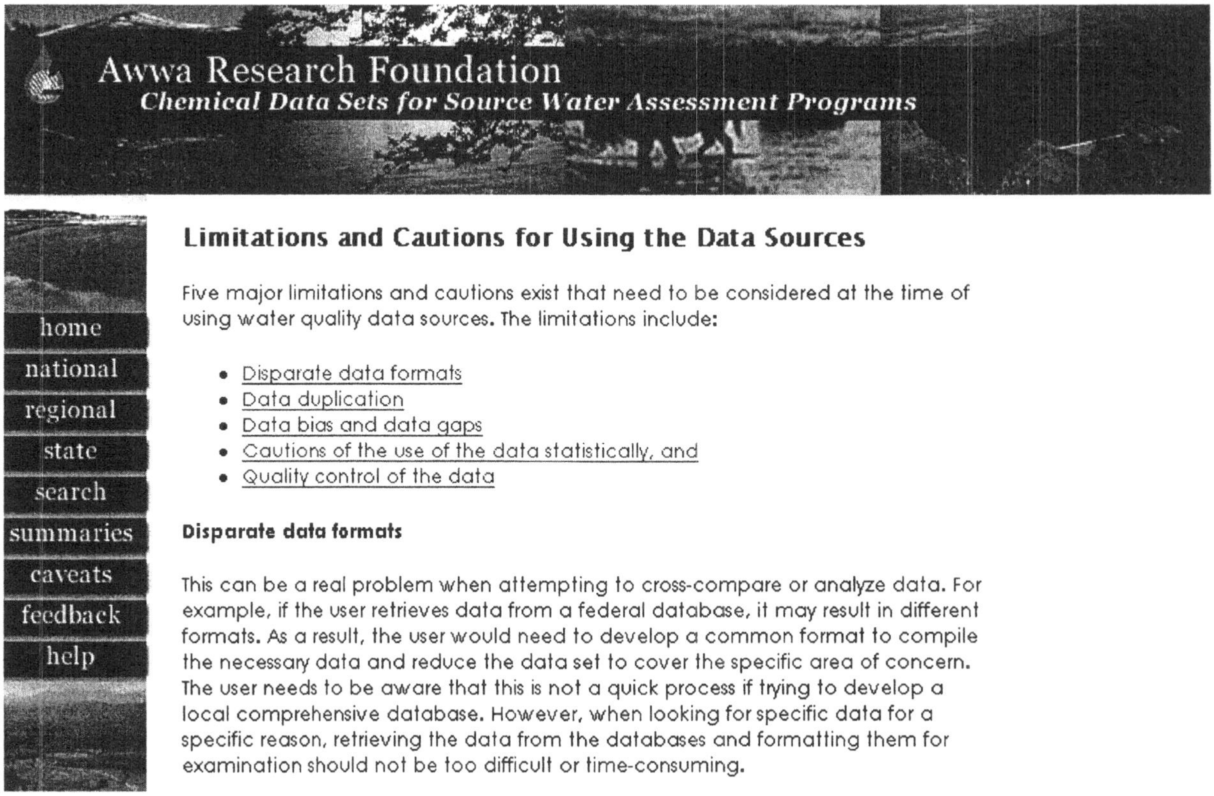
Awwa Research Foundation
Chemical Data Sets for Source Water Assessment Programs

home
national
regional
state
search
summaries
caveats
feedback
help

Limitations and Cautions for Using the Data Sources

Five major limitations and cautions exist that need to be considered at the time of using water quality data sources. The limitations include:

- Disparate data formats
- Data duplication
- Data bias and data gaps
- Cautions of the use of the data statistically, and
- Quality control of the data

Disparate data formats

This can be a real problem when attempting to cross-compare or analyze data. For example, if the user retrieves data from a federal database, it may result in different formats. As a result, the user would need to develop a common format to compile the necessary data and reduce the data set to cover the specific area of concern. The user needs to be aware that this is not a quick process if trying to develop a local comprehensive database. However, when looking for specific data for a specific reason, retrieving the data from the databases and formatting them for examination should not be too difficult or time-consuming.

Figure 3.10 Caveats web page

State Survey

The state survey (Figure 3.11) was and will be accessed by the participating states providing information on their data source and data sets. To reach the state survey after entering the site, the user has to click on "home" on the left navigation bar and then click "state survey." The state survey can only be accessed using the home page. Only authorized users may enter data on this survey. The user will be prompted for a "user ID" and "password" provided by TPMC via the feedback process (see below). Before new information is added to the existing database, it is reviewed for consistency and quality.

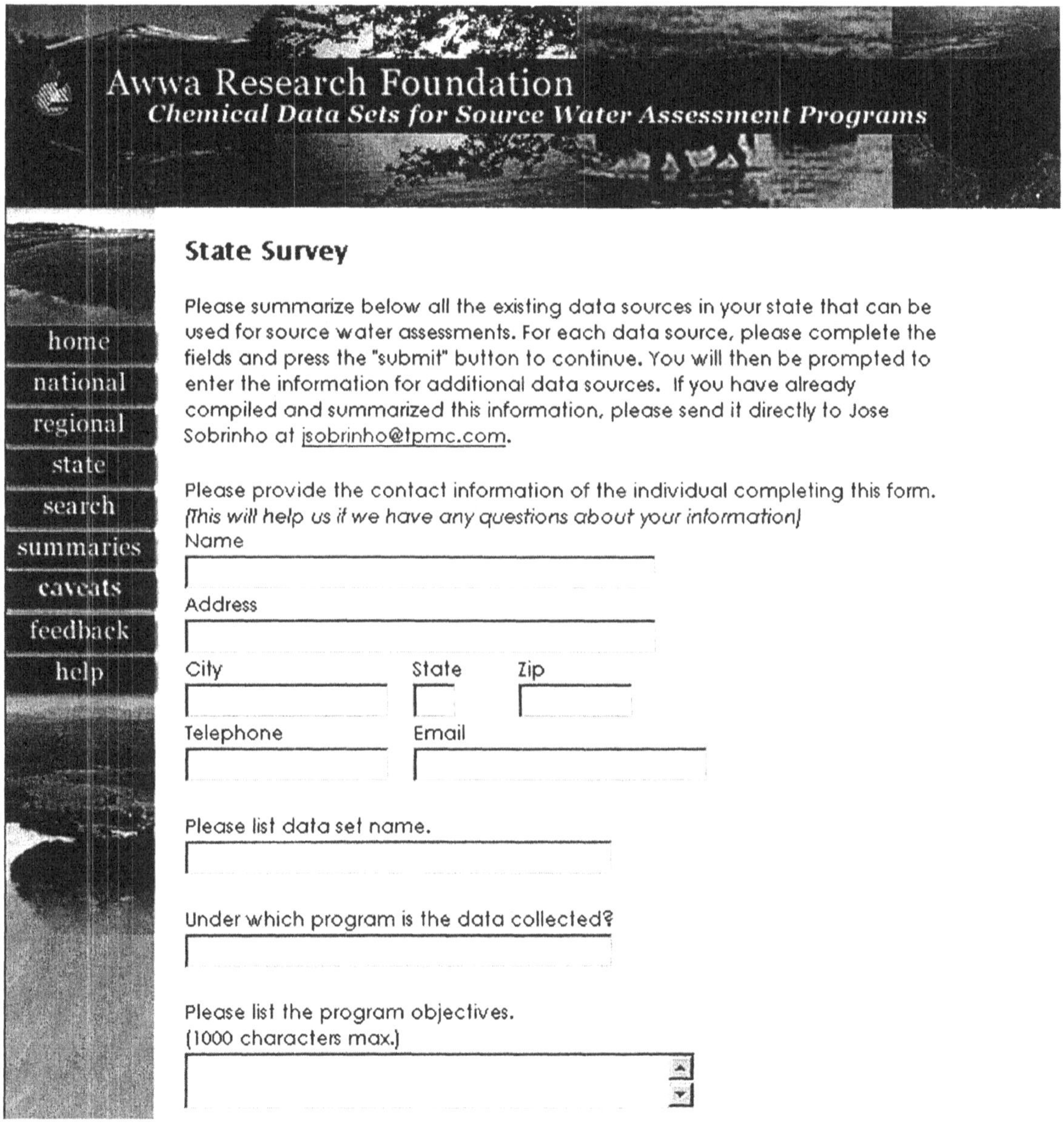

Figure 3.11 State survey web page

Help

The help page provides instructions regarding the use of and how to navigate through the website. In addition, a brief methodology for the data collection is also included in this page.

RESULTS

Table 3.1 presents a matrix for all data sets included in the source water quality database inventory. The matrix displays how many data sets contain specific types of data including:

- What is the geographic extent (national, regional, state, watershed, waterbody, or aquifer) of the data set?
- What time frame does the data set cover?
- What chemicals are included in the data set?
- Are sampling methods documented for the data set?
- Is a location or description of the sampling points included for the data set?
- Are the analytical methods and associated detection limits well documented for the data set?
- Were quality assurance and quality control procedures for sample collection and analysis performed for the data set?
- What is the overall assessment of the quality of the data based on the completeness of the data set?

This matrix is intended to provide an executive summary of each data set. Data are presented starting with national data sets followed by regional and state data sets. An "x" in the columns represent that the data set does include that information or meet that requirement. For the data sets researched by the Project Team the overall assessment of the metadata may need to be augmented by state agencies as full information was not available online. The Project Team used Best Professional Judgment and the available resources to make an educated decision about the metadata.

Table 3.1
Matrix of chemical occurrence data sets

Data set Name	Geographic Extent			Time Frame		Chemical Groups									Metadata						
	National	Regional	State	Start Date	End Date	General Physiochemical	Nutrients	Other Inorganics	Pesticides and SOCs	VOCs	Radionuclides	Particulates	Metals	General Organics	Sampling Area Description	Sampling Methods	Locational Sampling Points	Analytical Methods	Detection Limits	QA/QC Sample Collection	QA/QC Analysis
STORET Legacy Data Center (LDC)	x			1/1/60	12/31/98	x	x	x	x	x	x	x	x	x	x	x	x	x	x	x	x
Modernized STORET	x			1/1/99	Present	x	x	x	x	x	x	x	x	x	x	x	x	x	x	x	x
National Water Quality Assessment Program (NAWQA)	x			1/1/91	12/1/97		x	x	x	x	x		x		x	x	x	x			
National Stream Quality Accounting Network (NASQAN)	x			1/1/73	Present	x	x	x	x			x	x	x	x		x				
Nutrient Criteria	x						x								x	x	x	x	x	x	x
National Water Information System (NWIS)	x			1/1/60	1/1/99	x	x	x	x	x	x	x	x	x	x	x	x	x			
Safe Drinking Water Information System (SDWIS)	x			1/1/93	1/1/01	x	x	x	x	x	x	x	x	x	x		x	x	x	x	x
National Contaminant Occurrence Database (NCOD)	x			1/1/83	1/1/01	x	x	x	x	x	x	x	x	x	x	x	x	x	x	x	x
Kansas Stream Chemistry Database		x		7/5/67	12/31/01	x	x	x	x		x	x	x		x	x	x	x	x	x	x
Chesapeake Bay Program Water Quality Database		x		1/1/84	Present	x	x	x					x	x	x	x	x	x			
Chesapeake Bay Toxics Program Database		x		1/1/73	Present			x	x	x			x	x	x	x	x	x			
Alliance Citizen Monitoring Program Database		x		1/1/85	Present	x	x								x	x	x	x			
Arkansas STORET		x		1/170	Present	x	x	x	x	x		x	x	x			x		x		
New England Interstate Water Pollution Control Commission (NEIWPCC)		x		1/1/90	1/1/99	x	x								x	x	x	x	x	x	x
Occurrence and Distribution of Pesticides in Chesapeake Bay		x		1/1/76	1/1/94				x						x	x		x	x		

(continued)

Table 3.1 (continued)
Matrix of chemical occurrence data sets

Data set Name	Geographic Extent			Time Frame		Chemical Groups									Metadata						
	National	Regional	State	Start Date	End Date	General Physiochemical	Nutrients	Other Inorganics	Pesticides and SOCs	VOCs	Radionuclides	Particulates	Metals	General Organics	Sampling Area Description	Sampling Methods	Locational Sampling Points	Analytical Methods	Detection Limits	QA/QC Sample Collection	QA/QC Analysis
Groundwater Protection Council (GWPC) Environmental Information Suite (in development)		x				na	na	na	na	na	na	na	na	na							
Schuylkill and Delaware River Watershed Database		x		1/170	1/102	x	x	x	x	x	x	x	x	x	x		x		x		
Reservoir Monitoring Data			x (AL)	4/1/90	10/31/01	x	x									x	x	x	x	x	x
State of Alaska Public Water System Supervision (PWSS) Database			x (AK)	1/1/80	12/31/00		x	x	x	x	x	x	x	x	x		x	x	x		
Water Quality Division Database			x (AZ)			x		x	x	x		x			x					x	x
Malibu Creek Stream Team Database			x (CA)	11/1/98	Present	x	x					x			x	x	x	x			
Department of Water Resources Water Data Library			x (CA)			x		x	x			x		x				x	x		
Department of Pesticides Regulation's Surface Water Monitoring Database			x (CA)	1/1/97	Present	x	x	x					x		x	x	x	x	x		
Rivers of Colorado Water Watch Network Database			x (CO)	1/1/90	Present	x	x	x					x						x		
Delaware Public Water Supply Systems Database			x (DE)		1/28/02	x	x	x	x	x	x		x	x			x	x	x	x	x
Generalized Water Information System (GWIS3)			x (FL)	1/1/87	Present	x		x	x	x		x	x		x						
Radold			x (GA)								x				x						
Safe Drinking Water Information System/State			x (GA)			x	x	x	x	x		x	x	x	x				x	x	

(continued)

Table 3.1 (continued)
Matrix of chemical occurrence data sets

Data set Name	Geographic Extent			Time Frame		Chemical Groups									Metadata						
	National	Regional	State	Start Date	End Date	General Physiochemical	Nutrients	Other Inorganics	Pesticides and SOCs	VOCs	Radionuclides	Particulates	Metals	General Organics	Sampling Area Description	Sampling Methods	Locational Sampling Points	Analytical Methods	Detection Limits	QA/QC Sample Collection	QA/QC Analysis
Department of Health Water Quality			x (HI)	1/1/74	4/30/02		x	x	x	x	x		x		x			x	x		
Idaho Surface Water Quality Statewide Network			x (ID)	1/1/90	Present	x	x	x				x				x					
Illinois Source Water Assessment Database			x (IL)	1/1/85	Present		x	x	x	x			x			x	x	x		x	x
The Lower Illinois River Basin NAWQA Study Unit			x (IL)	1/1/84	1/1/91		x	x					x		x						
Radium in Groundwater from Public Supply Aquifers in Northern Illinois			x (IL)								x										
Drinking Water Compliance Database and SDWIS/State			x (IN)	1/1/93	12/31/01	x	x	x	x	x	x	x	x	x	x	x	x	x	x		
Iowa STORET			x (IA)	10/1/99	Present	x	x	x	x	x		x	x		x	x	x	x	x	x	x
Environmental Monitoring for Public Access and Community Tracking/Des Moines Database			x (IA)	6/1/99	Present	x	x	x	x			x	x								
Kansas Lake and Wetland Water Quality Database			x (KS)	5/1/75	12/31/01	x	x	x	x			x	x		x	x	x	x	x	x	x
Kansas Groundwater Quality Database			x (KS)	4/10/85	12/31/01	x	x	x	x	x	x		x		x	x	x	x	x	x	x
Compass			x (KY)	1/1/90	6/30/02	x	x	x	x	x		x	x	x	x	x	x	x	x	x	x
The Kentucky Lake Long-Term Monitoring Program Data set			x (KY)	7/1/88	Present	x	x	x				x			x						
The Lake Barkley Database			x (KY)	7/1/88	Present	x	x	x				x			x						
Ledbetter and Panther Watersheds Database			x (KY)	1/1/95	Present	x	x	x				x			x						

(continued)

Table 3.1 (continued)
Matrix of chemical occurrence data sets

Data set Name	Geographic Extent			Time Frame		Chemical Groups									Metadata						
	National	Regional	State	Start Date	End Date	General Physiochemical	Nutrients	Other Inorganics	Pesticides and SOCs	VOCs	Radionuclides	Particulates	Metals	General Organics	Sampling Area Description	Sampling Methods	Locational Sampling Points	Analytical Methods	Detection Limits	QA/QC Sample Collection	QA/QC Analysis
Public Water Supply Water Quality Data			x (ME)	1/1/80	1/1/02	x	x	x	x	x	x	x	x	x				x	x		x
Public Educational Access to Environmental Information in Maine (PEARL)			x (ME)	1/1/80	12/31/01	x	x								x	x	x				
Drinking Water Monitoring Data - SDWIS			x (MD)	1/1/93	1/1/01	x	x	x	x	x	x	x	x	x	x		x	x	x		
Maryland Department of Environment (MDE) / Department of Natural Resources (DNR) Core Trend Data			x (MD)	1/1/74	1/1/01	x	x	x					x	x	x		x				
Water Quality Testing System			x (MA)	4/1/93	2/11/02		x	x	x	x	x		x		x		x	x	x		
Site/Reportable Releases Look Up			x (MA)_					x	x	x			x	x	x	x	x		x		
Underground Storage Tank (UST) Registry			x (MA)							x				x	x					x	
Act 452 Part 201 Environmental Remediation Sites			x (MI)	6/1/90	12/15/01	x	x	x	x	x	x	x	x	x	x		x				
Leaking Underground Storage Tanks			x (MI)	1/1/80	12/31/01					x					x						
National Pollutant Discharge Elimination System (NPDES) Permit List			x (MI)	3/29/74	12/31/01	x	x	x	x	x	x	x	x	x	x		x				
WaterChem Database			x (MI)	1/1/93	Present	x	x	x	x	x	x	x	x	x	x						
Public Water Supply Supervision Program Compliance Monitoring			x (MN)	1/1/96	Present		x		x	x	x	x	x			x	x	x	x	x	x

(continued)

Table 3.1 (continued)
Matrix of chemical occurrence data sets

Data set Name	Geographic Extent			Time Frame		Chemical Groups									Metadata						
	National	Regional	State	Start Date	End Date	General Physiochemical	Nutrients	Other Inorganics	Pesticides and SOCs	VOCs	Radionuclides	Particulates	Metals	General Organics	Sampling Area Description	Sampling Methods	Locational Sampling Points	Analytical Methods	Detection Limits	QA/QC Sample Collection	QA/QC Analysis
Nitrate Nitrogen Analyses from Private Water Supply Wells			x (MN)	1/1/90	Present		x										x	x	x	x	x
Groundwater Monitoring and Assessment Program (GWMAP) Database			x (MN)	1/1/91	1/1/96	x		x		x			x	x	x		x				
Statewide Surface Water Monitoring for Nutrients and Pesticides			x (MN)	1/1/92	Present		x		x												
Source Water Assessment Program (SWAP) Geographic Information System (GIS) (in development)			x (MS)			na	na	na	na	na	na	na	na	na			x				
Missouri Stream Team Database			x (MO)	1/1/93	Present	x	x						x					x	x		
Montana Natural Resource Information System			x (MT)			x	x						x		x	x	x	x			
Quality-Assessed Agrichemical Contaminant Database for Nebraska Groundwater			x (NE)	1/1/74	1/1/02				x								x	x	x	x	x
State of Nevada Surface Water Monitoring Network - Streams			x (NV)	1/1/90	Present	x	x	x				x	x		x		x				
State of Nevada Surface Water Monitoring Network - Lakes and Reservoirs (in development)			x (NV)	1/1/90	Present	na	na	na	na	na	na	na	na	na	x		x				
Watershed Surface Water Quality Data			x (NH)			x	x	x				x	x	x							
New Jersey United States Geological Survey			x (NJ)						x	x	x				x						
Water Resources Data – Ground/Surface Water Data			x (ND)	1/1/70	Present	x		x							x	x	x	x	x		

(continued)

Table 3.1 (continued)
Matrix of chemical occurrence data sets

Data set Name	Geographic Extent			Time Frame		Chemical Groups									Metadata						
	National	Regional	State	Start Date	End Date	General Physiochemical	Nutrients	Other Inorganics	Pesticides and SOCs	VOCs	Radionuclides	Particulates	Metals	General Organics	Sampling Area Description	Sampling Methods	Locational Sampling Points	Analytical Methods	Detection Limits	QA/QC Sample Collection	QA/QC Analysis
Miami Conservancy District – Groundwater Quality Data			x (OH)			x		x	x	x						x	x	x	x	x	x
Oklahoma Environmental Information Management Suite (EIMS)			x (OK)				x										x				
Oregon Safe Water System: Chemical Detections, Burns Water Department			x (OR)					x	x	x	x				x	x	x	x	x	x	x
Laboratory Analytical Storage and Retrieval (LASAR) Database			x (OR)	1/1/49	Present	x		x	x				x	x	x	x	x	x	x	x	x
Oregon Water Quality Index			x (OR)	1/1/85	Present	x	x							x	x						
Susquehanna River Basin Commission (SRBC) Nutrient Assessment Program			x (PA)			x	x								x		x	x	x		
Pennsylvania Spatial Data Access (PASDA) – The Pennsylvania Geospatial Data Clearinghouse			x (PA)			na	na	na	na	na	na	na	na	na	x						
Rhode Island Public Water Inventory			x (RI)	1/1/70	4/29/02	x	x	x	x	x	x	x	x	x	x	x	x	x	x	x	x
South Carolina Ambient Ground Water Quality Monitoring Network			x (SC)	1/1/87	Present	x	x	x				x	x	x	x	x	x				
South Carolina Groundwater Contamination Inventory			x (SC)	1/1/87	Present		x		x	x	x		x		x						
Source Water Assessment Program (SWAP) Potential Source of Contamination (PSOC) Data set			x (TX)	10/1/86	5/31/02	x	x	x	x	x	x		x	x	x		x			x	x

(continued)

Table 3.1 (continued)
Matrix of chemical occurrence data sets

Data set Name	Geographic Extent			Time Frame		Chemical Groups									Metadata						
	National	Regional	State	Start Date	End Date	General Physiochemical	Nutrients	Other Inorganics	Pesticides and SOCs	VOCs	Radionuclides	Particulates	Metals	General Organics	Sampling Area Description	Sampling Methods	Locational Sampling Points	Analytical Methods	Detection Limits	QA/QC Sample Collection	QA/QC Analysis
Texas Water Development Board (TWDB) Groundwater Database (GWDB)			x (TX)	1/1/60	Present	x	x						x			x	x	x	x	x	x
Texas Natural Resources Conservation Commission (TNRCC) – Sampling Data Query, Surface Water Quality Monitoring			x (TX)	1/1/93	Present	na	na	na	na	na	na	na	na	na		x		x	x		
Underground Storage Tanks (UST), Leaking Underground Storage Tanks (LUST), Brownfields, Superfund, Toxics Release Inventory (TRI), Voluntary Clean Program (VCP)			x (UT)	1/1/02	2/19/02	na	na	na	na	na	na	na	na	na		x					
Statewide Geographic Information Database			x (UT)	1/1/190	2/20/02	na	na	na	na	na	na	na	na	na		x					
Water Supply Division (WSD) Inventory			x (VT)	1/1/89	3/13/02	x	x	x	x	x	x	x	x		x				x		
Environmental Assessment Program – River Stream and Water Quality Monitoring			x (WA)	1/1/59	1/1/01	x	x					x			x	x	x	x	x		x
Pesticides in Public Supply Wells of Washington State			x (WA)	1/1/95	1/1/98		x		x						x						
West Virginia Source Water Assessment and Wellhead Protection Program SWAP data online (in development)			x (WV)			na	na	na	na	na	na	na	na	na							
Drinking Water System			x (WI)	5/3/92	5/3/02	x	x	x	x	x	x	x	x	x	x		x		x		
Statewide Water Resources Data Inventory			x (WY)			x	x	x	x	x	x	x	x	x	x		x				

Empty Time frame =start/end dates not available; Blank Metadata field = "not sure"; highlighted = "no"; "na" = not available

CHAPTER 4
UTILITY REVIEW OF THE PROJECT WEBSITE

BACKGROUND

The Philadelphia Water Department (PWD) and Water Revenue Bureau serve the Greater Philadelphia region by providing integrated water, wastewater, and storm water services. The utility's primary mission is to plan for, operate, and maintain both the infrastructure and the organization necessary to purvey high quality drinking water, to provide an adequate and reliable water supply for all households, commercial, and community needs, and to sustain and enhance the region's watersheds and quality of life by managing wastewater and storm water effectively.

Whether PWD is providing safe water for residents to drink, supplying water for industries to manufacture goods in Philadelphia, or protecting the region's water resources, serving their customers with quality has been a commitment of the PWD throughout the city's nearly 200-year history. Their commitment to collecting storm water and treating wastewater is a commitment to the health and vitality of the Delaware and Schuylkill Rivers, countless creeks and streams, and the people of the region who enjoy and depend on these waterways.

The PWD's Source Water Quality Database is a compilation of regional water quality information compiled from federal, state, and local databases. This includes data from STORET, USGS, water suppliers, volunteers, U.S. Army Corps of Engineers, consultants, watershed organizations, public health agencies, schools and universities, and industrial dischargers. The data ranges from 1970 to 2002, but the bulk of the data is between 1990 and 2000. This data includes specific locational/geographical components added by PWD. In addition, every location has a latitude/longitude and a stream/river tributary mile associated with it. The sites are graphically represented on computerized maps to aid in spatial comparisons as well as investigations that can link water quality in a given area to particular point sources, land uses, geographical formations, or soil types. This linking of water quality data with watershed characteristics via geographical information systems (GIS) creates a powerful tool to identifying the sources of various water quality challenges in the watershed. In addition, data gaps are identified to determine where there is a lack of temporal or spatial coverage in sampling data as well as the need to add or remove parameters from various sampling initiatives. This data is also compared with geographical representations of impaired stream information (303d TMDL).

PWD structures their approach to performing SWAPs by beginning with a section on the general watershed in review, starting with an introduction, background and history of the watershed, physiography and soils, hydrology, and land use. Then they move to a summary of past studies. A good starting point is to review the numerous reports developed by governmental agencies, watershed organizations, educational institutions, and citizen groups. Identification of universal water quality issues follows, including: long-term trends, changes in water quality over time, analysis of waterbody impairments, and current and future needs of watershed monitoring. Once all of the background is completed an inventory of potential sources of contamination, point sources, and inventory characterization is completed. Identification of restoration efforts, public participation (e.g., advisory groups and public meetings) and general recommendations for the watershed complete the SWAP.

PWD also recommends documentation of general issues which may be raised by the public when performing a SWAP, including:

- What is the source of the drinking water?
- Water quality and treatment information
- Evaluation of significant sources of contamination
- Ongoing source water protection activities
- Source water protection needs
- How can the public obtain more information?
- How can the public become more involved in protecting the water supply?

SOURCE WATER QUALITY DATABASE WEBSITE REVIEW

PWD reviewed the website and noted the excellent job at making this a good starting point for users (e.g., manager at utilities, drinking water administrators) to visit. Of specific interest was the information on national, regional and state data sets, and the links to the state SWAP programs, as it is sometimes difficult and time consuming to locate SWAP information from other states, which can be a major issue in data retrieval. This data inventory website is the first step to solve this problem. The Project Team is hopeful that once the website gains recognition from the scientific community, more databases will become available, and it will be more valuable to everyone.

PWD felt that the website would have helped them to identify, access, and understand national data sets. However, PWD suggested the website should provide some basic cautionary information which is critical when working with data from disparate data sets to help the user understand how to assimilate, compile, format, and structure requests from these databases. To accommodate this, a "Caveats" page was included in the website listing the limitations related to data use and provides recommendations to address the given limitations. The PWD suggestions and recommendations to follow when performing data mining from disparate data sources are detailed in the following two sections based on lessons learned from their own data mining experience when collecting data for their SWAP.

LIMITATIONS AND DATA CAUTIONS

Five major data cautions were recognized during the PWD review and critique on the project website, including: disparate data formats, data duplication, data bias and data gaps, cautions of the use of the data statistically, and quality control of the data.

Disparate Data Formats

This can be a real problem when attempting to cross compare or analyze data. For example, if the user retrieves data from a federal database it may result in different formats. As a result, the user would need to develop a common format to compile the necessary data and then reduce the data set to cover the specific area of concern. The user needs to be aware that this is not a quick process if they are trying to develop a local comprehensive database. However, if they are looking for specific data for a specific reason, retrieving the data from the databases and formatting them for examination should not be too unwieldy or time consuming.

Data Duplication

Data may be submitted to national databases (e.g., STORET) from state agencies and still also be contained in the state agency database. If the user goes to the national databases and downloads data, then does the same with state and local they risk data duplication. The user should be aware that the potential exists for data duplication.

Data Bias and Data Gaps

Since a study can be based on existing data sets, the nature of the data and an understanding of the sampling plans and sampling objectives need to be explored. To do this by documenting each data set would be difficult. The types of problems, which might be included in the underlying data, include biases in sampling due to contagious sampling. For example, biases may arise if a problem was identified within a system and then the system was sampled extensively near the problem point but was not sampled away from the problem or in systems that did not have problems. Inclusion of all this data can have serious effects on all summary statistics and could result in criteria that represent severely impacted sites. Other examples of biases based on grouping data from many studies include pooling data from intensive short-term studies with low intensity long-term studies. An example would be a study, which sampled fixed stations once each season for 25 years versus a study that sampled each station daily for one year. Pooling these data sets can result in unintended biases in distributional parameters and characteristics.

Use of the Data Statistically Should be Approached With Caution

Once states compile data sources they must take into account several cautions regarding data bias during statistical analysis as well. The program under which the data were collected and the purpose of data collection are of critical importance. The objectives of a monitoring program or data collection strongly dictate the sampling program that is employed. Taken alone the data can be analyzed using standard statistical methodology. However, when data collected from different programs are merged to gather the differences in the statistical sampling, the analysis and the aggregation of the data can be critical. Not taking these things into account can result in erroneous conclusions. As a hypothetical example, imagine a very important waterbody in a large community that supports recreation, education, industry and potable water uses. If the waterbody is very large, then it is likely that many users will be monitoring the water quality. Some of the monitoring will be based on regulatory compliance and therefore will have strict periodicity, analytical methods and purposes. A local university that is interested at understanding ecosystem conditions might undertake regional monitoring. A beach might be monitoring for basic water parameters and microbiology. At the same time a scout camp on the shore of the waterbody might have science projects conducted by the scouts. In all cases the monitoring has a specific objective and planned sampling in both space and time. The analyses range from well meaning citizens through professional, certified laboratories. An analyst aggregating these results in an effort to understand the water quality of the waterbody could take the hundreds of measurements made with uncalibrated instruments and average them with the single quarterly measurement made for compliance monitoring and end up with an erroneous result. This may happen because the abundant data from some sources (e.g., citizen groups) far out shadows the quarterly data measured by the professional laboratory. Similarly the abundant compliance data will be specific to the intake or the outflow from a plant or industry. These data are highly biased. If combined with a research program's randomized sampling plan that covers the entire waterbody then the compliance data will overwhelm the sparse but highly valued randomized data collection. Giving these results the same weight will result in an incorrect inference about the overall quality of the waterbody.

Quality Control of the Data

Similarly the level of quality assurance performed is critical as illustrated in the example above. The amount of quality control will be dictated by the questions being addressed and the budgets available. Another useful example is related to the dissolved oxygen concentration in a waterbody. If one researcher is interested in anoxic conditions then the precise concentrations are not as important as the presence or absence of some oxygen. Therefore the calibration of the instrument or the titration methodology may not be as critical as a researcher who is looking at hypoxic conditions for which highly sensitive calibration is necessary. Mixing these two types of data for a given waterbody or sampling station could result in results that are not correct or defensible.

RECOMMENDATIONS FOR USING THE WATER QUALITY DATA SOURCES

The data contained in the source water quality database inventory is a critical resource that needs to be approached with caution. Often users know data are available but they are unsure of how to access it, avoid data duplication, data bias, and how to analyze data from disparate data sets. Some basic recommendations to assist users with the data mining and data management process along with approaches for managing the data limitations are discussed below.

The Data Mining Process

Before the user begins mining water quality data, they need to evaluate what they want to answer and what will be the ultimate use for the data. If the user only has one or two questions and believes they will not be revisiting the data in the future to answer more questions, then simply downloading the locations, time periods, parameters, and compiling the data in something as simple as Microsoft Excel may be the most efficient process.

However, if the user is planning to conduct extensive exploratory analyses of the data set to examine trends, compare locations, or conduct statistical analyses, it is recommended that careful thought, planning, and resources be allocated prior to data mining. This includes examining all the databases and data sources and comparing their formats. Then a determination must be made about which fields from those databases need to be retained (some databases can have over 100 fields describing the particulars on sampling or analytical methods for a given sample). Based on that information, a database can be designed that will hold the data, but also have adequate capabilities to retrieve the data in automated queries or other features. For example, linking the data to GIS mapping enabling the user to a click on a location and query specific information about the data would be useful. Usually Microsoft Access software is sufficient for most data needs as Microsoft Access is fairly compatible with GIS mapping programs.

Protocol for Database Development That can Assist With the Issue of Data From Disparate Format

Developing a comprehensive database requires some data access skills and protocols. Downloading data is usually a straightforward task, however, organizing and formatting the data into a framework that is useful for retrieval requires some energy and repetitive actions with attention to detail. An organization's staff may change throughout this process; therefore it is imperative to develop protocols on exactly how the data is formatted for the database and how specific additional pieces of information (or fields) are added to the data to make it relevant to your database. If the user is downloading and formatting a lot of data, developing automated codes for formatting or uploading data can save significant amounts of time and reduce operator error.

Advice and/or Constraints for Reviewing Data Sets

When obtaining data from the various databases, there are a number of ways to proceed to collect them. First, data can be downloaded and then the user can format it and sort through it later. This is usually the most time intensive of the choices. Second, the user can examine which locations have sufficient data to meet the data objectives (i.e., number of observations, time period, geographical locations, etc.) and then download those specific files. For example, if the user is interested in nutrient data for the Delaware River between 1990 and 1999 the data download can be narrowed from hundreds of locations to less than a dozen without a lot of work. Then it can be determined if the data are sufficient for the specific areas of interest in the watershed or if the number of observations is sufficient. In some cases, even these requests may be too large and unwieldy for the computers or Internet to handle downloads in a timely manner. Therefore, it may be necessary to consider dividing requests and downloads into smaller pieces to recombine at a later time. This is more time consuming and produces greater chances for error and duplication; therefore careful precautions must be followed.

Customize Information From the Data Sets

Developing some background information for each database would be useful, as it can be difficult to constantly manipulate the national databases and apply filters to the data. For example, PWD downloaded and formatted as much as possible into their database so it possesses some unique attributes that other databases may be lacking. PWD obtained the latitude and longitudes for the site and assigned a table with the watershed, stream, subwatershed to each of the sites in an effort to provide querying capabilities at various levels. They also developed queries by date and by parameter and can have the data automatically come up in formats for long-term trend. Also, the data source and organization is provided to enable a number of ways for filtering data, as certain data sets may bias the statistical analysis. For example, if the user only wants data for a specific waterbody and time period, but does not want data collected by volunteer organizations, and only wants sites located between certain stream or river miles, the database is flexible to accommodate such a request. PWD found that many data sets are lacking strong geographical components, so they suggest using GIS skills to enhance each data set significantly.

Conduct Some Basic Exploratory Analyses on the Different Data Sets to Assess any Potential Data Bias/Data Gaps and/or Data Duplication

Exploratory analyses can result in an understanding of these potential problems and possible sources of bias in the distributions. Based on the exploratory analyses, transformation of the data and aggregations can be performed which will identify appropriate statistical methodologies for evaluating the hypotheses. The exploratory analyses will focus on the development of distributions for all the data collected as well as for some categorized data. The development of the categories can be based statistically or on established expectations. This approach should also help with the limitations of using the data statistically. Since data may be based on various sampling techniques it is imperative to research this and begin with the lowest level of analyses to document any similarities or major dissimilarities between the data sets.

CHAPTER 5
SUMMARY AND CONCLUSIONS

When developing initiatives for source water assessment and protection (e.g., SWAPs) it is necessary to have strong scientific data to justify the need for protection or restoration of a waterbody. Since water quality monitoring is usually being conducted by numerous organizations at any given time without coordination or sharing, this data requires compilation into a cohesive database or information system for use and analysis by water suppliers and other stakeholders. This provides not only the current source water quality profile, but past and future water quality trends that can indicate sources and locations of contamination as well as identification of data gaps in ongoing monitoring efforts by the water supplier or other agencies. A water supplier can then use this information to develop strategies to address pollution, educate and influence local organizations or regulatory agencies to act to reduce or mitigate a source of pollution, or to plan for the operational and capital construction resources to address future source water quality changes in relation to future regulatory requirements and customer goals. In cases where a compound of potential concern is identified in national studies, other indicator parameters can be used to provide preliminary information to managers and customers about the likelihood that it will be a local issue.

The source water quality database website, the tool specially designed to assist water suppliers and other stakeholders, stores information on twelve major national databases, nine regional and seventy-one state data sets. The intent of the source water quality database website is to help everyone involved in meeting the requirements for source water assessment programs (e.g., managers at utilities, drinking water administrators, consultants) to fulfill compliance. Data resources were pooled from all over the country including national, regional, and state databases/data sets into one easily accessible website.

Data sources include results from a variety of programs managed by state pollution control agencies, Indian tribes, and local governments such as city and county environmental offices. In addition, federal agencies such as USGS and private entities (e.g., universities, watershed associations, environmental groups, and industries) were also queried where possible.

As a result of the differences in quality of the data managed by those programs, standards used for sample collection and analytical work, and spatial data coverage may vary widely, the data sets and databases were inventoried and reviewed based on a variety of descriptors.

Data quality of each source was reviewed to ensure reliability of the inventory. State SWAP contacts and links to state SWAP websites are also included. In summary, this website is designed to help utility personnel, state drinking water administrators, and source water coordinators address requirements for SWAPs.

CHAPTER 6
RELEVANCE TO THE DRINKING WATER COMMUNITY

Chemical water quality information is a critical resource useful to characterize, protect and preserve the condition and uses and also water quality improvements attained in the waters throughout the nation. An enormous amount of chemical data and information are currently collected, stored and managed by state pollution control agencies, Indian tribes, and many local governments, such as city and county environmental offices. Federal agencies such as USGS, through the NAWQA program, and the US Army Corps of Engineers are involved in chemical monitoring as well. Private entities such as water suppliers, universities, watershed associations, environmental groups, and industries also conduct water quality monitoring.

When chemical data, collected by these various agencies and private entities, are needed by SWAP or TMDL programs, data cannot necessarily be easily retrieved. Typically, only a small audience can easily access the data. Without this data, it is not known where pollution problems exist, where pollution control strategies should be focused, or where or how much progress has been made.

The development and implementation of an inventory of data for use in developing SWAPs has numerous practical benefits to the drinking water community. The product of this research has extensive application to the drinking water and regulatory communities for assessing source waters, enriching the ability to assemble and evaluate all existing and readily available water quality related data and information. Specifically, this project documented chemical data stored in a variety of databases and provided a mechanism to access it and the associated metadata. The online tool, Source Water Quality Database, provides information on water quality data related to chemical occurrence stored in national, regional and states data sets. The project results also can help organizations that monitor and use chemical water information for the:

- Characterization of source waters and identification of changes or trends in water quality over time;
- Identification of existing or emerging water quality problems resulting in increased public health protection against chemical exposures;
- Design of specific pollution prevention or remediation programs; and
- Determination of whether program goals such as compliance with pollution regulations or implementation of effective pollution control actions are being met design monitoring programs.

APPENDIX A
STATE CONTACT LIST (Last Updated: May 2002)

State	Contact Name and address	Telephone	E-mail
Alabama	Lynn Sisk Chief of the Water Quality Branch Alabama Department of Environmental Management 1400 Coliseum Boulevard PO Box 301463 Montgomery, AL 36130-1463	334.271.7826	ls@adem.state.al.us
Alaska	James Weise Program Manager/Drinking Water and Wastewater Program Alaska Department of Environmental Conservation Division of Environmental Health 555 Cordova Street Anchorage, Ak 99501-2617	907.269.7647	James_Weise@envircon.state.ak.us
Arizona	Jeff Stuck Drinking Water Section Manager Arizona Department of Environmental Quality 3033 N. Central Ave., MO248A Phoenix, AZ 85012	602.207.4617	stuck.jeff@ev.state.az.us
Arkansas	Ginger Tatom SWAP Supervisor Arkansas Department of Health Division of Engineering (MS37) 4815 West Markham St Little Rock, Arkansas 72205-3867	501.661.2623	gtatom@healthyarkansas.com
California	Paul Collins Chief of the Data System Support Unit California Department of Health Services Division of Drinking Water and Environmental Management 601 North 7th Street, MS 92 PO Box 942732 Sacramento, CA 94234-7320	916.324.2320	pcollins@dhs.ca.gov
Colorado	Gary Karst SWAP Coordinator Colorado Department of Public Health and Environment Water Quality Control Division 4300 Cherry Creek Drive South Denver, CO 80246-1530	303.692.3579	gary.karst@state.co.us

(continued)

STATE CONTACT LIST, continued

State	Contact Name and address	Telephone	E-mail
Connecticut	Lori Mathieu SWAP Coordinator State of CT Department of Public Health Water Supplies Section 410 Capitol Ave, MS#51WAT Hartford, CT 06134-0308	860.509.7343	lori.mathieu@po.state.ct.us
Delaware	John T. Barndt, P.G. DE Department of Natural Resources and Environmental Control Division of Water Resources Water Supply Section PO Box 1401 Dover, DE 19903	302.739.4793	jbarndt@dnrec.state.de.us
Florida	Allan Stodghill, P.G. Florida Department of Environmental Protection 2600 Blair Stone Rd MS # 3575 Tallahassee, FL 32399-2400	850.921.9429	allan.stodghill@dep.state.fl.us
Georgia	Brad Addison Program Manager Georgia Environmental Protection Division Drinking Water Compliance Program 205 Butler Street SE, Suite 1362 Atlanta, GA 30334	404.656.4807	brad_addison@dnr.state.ga.us
Hawaii	Bill Wong Chief of the Safe Drinking Water Branch Hawaii Department of Health 919 Ala Moana Blvd., Room 308 Honolulu, Hi 96814	808.586.4258	wwong@eha.health.state.hi.us
Idaho	David Risley Source Water Assessment Program Lead Idaho Department of Environmental Quality 1410 North Hilton, Boise, ID 83706	208.373.0274	drisley@deq.state.id.us
Illinois	Rick Cobb Manager Illinois Environmental Protection Agency Division of Public Water Supplies Groundwater Section PO Box 19276 Springfield, IL 62794-9276	217.785.4787	rick.cobb@epa.state.il.us

(continued)

STATE CONTACT LIST, continued

State	Contact Name and address	Telephone	E-mail
Indiana	Al Lao Indiana Department of Environmental Management Drinking Water Branch PO Box 7148 Indianapolis, IN 46206-7148	317.308.3283	ALAO@dem.state.in.us
Iowa	Michael K. Anderson SWAP Program Coordinator Iowa Department of Natural Resources Drinking Water Section 401 Southwest 7th Street Suite M Des Moines, Iowa 50309	515.725.0336	michael.anderson@dnr.state.ia.us
Kansas	Robert Angelo Kansas Department of Health and Environment Division of Environment Bureau of Environmental Field Services 1000 SW Jackson St., Suite 430 Topeka, KS 66612	785.296.8027	bangelo@kdhe.state.ks.us
Kentucky	Lee Colten Manager for Division of Water Kentucky Department for Environmental Protection 14 Reilly Road Frankfort, Kentucky 40601	502.564.3410	lee.colten@mail.state.ky.us
Louisiana	Les Lemon Louisiana Department of Public Health Engineering 6867 Bluebonnet Blvd. Baton Rouge, LA 70810	225.765.5054	llemon@dhh.state.la.us
Maine	Andrews Tolman Bureau of Health, Division of Health Engineering Maine Drinking Water Program 157 Capital Street Augusta, ME 04333-0010	207.287.6196	andrews.L.Tolman@state.me.us
Maryland	John Grace Maryland Department of the Environment 2500 Broening Highway Baltimore, MD 21224	410.631.3702	jgrace@mde.state.md.us

(continued)

STATE CONTACT LIST, continued

State	Contact Name and address	Telephone	E-mail
Massachusetts	Damon Guterman MA Department of Environment Protection Drinking Water Program One Winter Street 6th Floor Boston, MA 02108	617.574.6811	damon.guterman@state.ma.us
Michigan	Bradley B. Brogren, P.E. Michigan Department of Environmental Quality Drinking Water and Radiological Protection Division 525 West Allegan, PO Box 30630 Lansing, MI 48909-8130	517.241.1361	brogrenb@michigan.gov
Minnesota	Bruce Olsen Source Water Protection Supervisor Minnesota Department of Health 121 East Seven Place Suite 203 St. Paul, MN 55101	651.215.0800	bruce.olsen@health.state.mn.us
Mississippi	Jamie Crawford Chief of the Ground Water Planning Branch Mississippi Department of Environmental Quality PO Box 10385 Jackson, MS 39289-0385	601.961.5354	jamie_crawford@deq.state.ms.us
Missouri	G. Lawson Penny Public Drinking Water Program Missouri Department of Natural Resources PO Box 176 Jefferson City, MO 65102	573.526.5449	
Montana	Joe Meek SWAP Section, Pollution Prevention Bureau MT Dept. Of Env. Quality Metcalf Building, Box 200901 Helena, MT 59620-0901	406.444.4806	jmeek@mt.gov
Nebraska	Marty Link NE Dept of Environmental Quality PO Box 98922, State House Stn. Lincoln, NE 68509-8922	402.471.4270	marty.link@ndeq.state.ne.us
North Carolina	Robert Midgette Public Water Supply Section NC Dept. of Environment and Natural Resources 1634 Mail Service Center Raleigh, NC 27699-1634	919.733.2321	Robert.Midgette@ncmail.net

(continued)

STATE CONTACT LIST, continued

State	Contact Name and address	Telephone	E-mail
North Dakota	Bill Gunnerson Ground Water Protection Program Division of Water Quality P.O. Box 5520 1200 Missouri Ave. Bismarck, ND 58506	701.328.5241	bgunners@state.nd.us
Nevada	Jim Balderson State Health Division Bureau of Health Prot. Serv. 1179 Fairview Dr., Ste. 101 Carson City, NV 89701-5405	775687.4754 x228	bhpsinfo@govmail.state.nv.us
New Hampshire	Paul Susca New Hampshire Dept. of Env. Services Water Supply Engineering Bureau P.O. Box 95 Concord, NH 03302	603.271.7061	psusca@des.state.nh.us
New Jersey	Sandy Krietzman NJ Dept. Env. Protection Bureau of Safe Drinking Water, CN 426 401 E. State St. Trenton, NJ 08625-0426	609.292.5550	skrietzm@dep.state.nj.us
New Mexico	John Gillentine Drinking Water Bureau NM Environment Dept. 525 Camino De los Marquez, Suite 4 Santa Fe, NM 87505	505-827.1400 x1043	john_gillentine@nmenv.state.nm.us
New York	Ron Entringer New York State Dept. of Health Bureau of Public Water Supply Protection Flanigan Square 547 River Street, RM 400, 4th floor Troy, NY 12180-2216	518.402.7713	rae01@health.state.ny.us
Ohio	Barb Lubberger Division of Drinking and Ground Water OH Environmental Prot. Agency P.O. Box 1049 Columbus, OH 43216-1049	614.644.2752	barb.lubberger@epa.state.oh.us
Oklahoma	Rance Shields Water Quality Division OK Dept. of Environmental Quality PO Box 1677 Oklahoma City, OK 73101	405.702.8120	shields.rance@deqmail.state.ok.us

(continued)

STATE CONTACT LIST, continued

State	Contact Name and address	Telephone	E-mail
Oregon	Sheree Stewart Drinking Water Protection Prog. OR Dept. of Environmental Quality 811 SW 6th Avenue Portland, OR 97204-1390	503.229.5413	stewart.sheree@deq.state.or.us
Pennsylvania	Christopher S. Crockett Philadelphia Water Department 1101 Market St. Philadelphia, PA 19107	215.685.6234	chris.crockett@phila.gov
Rhode Island	Clay Commons RI Dept. of Health Office of Drinking Water Quality 3 Capital Hill Providence, RI 02908-5097	401.222.6867	clayc@doh.state.ri.us
South Carolina	Rob Devlin Bureau of Water, Ground Water Management Section SC Dept. of Health and Env. Control 2600 Bull St. Columbia, SC 29201-1708	803.898.3798	devlinrj@dhec.state.sc.us
South Dakota	Anita Yan SD DENR Joe Foss Building 523 East Capitol Pierre, SD 57501-3181	605.773.3296	anita.yan@state.sd.us
Tennessee	Tom Moss Ground Water Management Section Division of Water Supply Dept. of Environment and Conservation 401 Church Street Nashville, TN 37243-1549	615.532.0170	tom.moss@state.tn.us
Texas	David Terry Public Drinking Water Section (MC-155) TX Natural Resource Conservation Comm. P.O. Box 13087 Austin TX, 78711-3087	512.239.4755	dterry@tnrcc.state.tx.us
Utah	Sumner Newman UT Dept. of Environmental Quality Division of Drinking Water P.O. Box 144830 150 North 1950 West Salt Lake City, UT 84114-4830	801.536.4195	snewman@deq.state.ut.us

(continued)

STATE CONTACT LIST, continued

State	Contact Name and address	Telephone	E-mail
Vermont	Elizabeth Hunt Water Supply Division Dept. Of Environmental Conservation 103 S. Main Street Waterbury, VT 05671-0403	802.241.3409	elizh@dec.anr.state.vt.us
Virginia	Christopher D. Adkins Division of Drinking Water 1500 E. Main Street, Room 109 Richmond, VA 23219	804.786.5568	cadkins@vdh.state.va.us
Washington	David Jennings Div. of Drinking Water, Dept. of Health P.O. Box 47849 Olympia, WA 98504-7849	360.236.3149	david.jennings@doh.wa.gov
West Virginia	Bill Toomey West Virginia Dept. of Health Environmental Engineering Division 815 Quarrier Street, Suite 418 Charleston, WV 25301	304.558.2981	wtoomey@wvdhhr.org
Wisconsin	Jeff Helmuth Wisconsin Dept of Natural Resources Bureau of Drinking Water and Groundwater PO Box 7921 Madison, WI 53707-7921	608.266.5234	helmuj@dnr.state.wi.us
Wyoming	Kim Parker WY Dept. of Environmental Quality Water Quality Division Herschler Building 4th West 122 West 25th Street Cheyenne, WY 82002	307.777.7343	kparke@state.wy.us

APPENDIX B
STATE DATA REQUEST LETTER

Ms. Barb Lubberger
Division of Drinking and Ground Water
OH Environmental Prot. Agency
P.O. Box 1049
Columbus, OH 43216-1049

Dear Ms. Lubberger:

We would appreciate your assistance with the ***"Chemical Occurrence Data sets for Source Water Assessment"*** project, which is funded by the American Water Works Association Research Foundation (AwwaRF). Background information on the project is attached. The objective of this project is to identify, list, and characterize chemical occurrence databases and data sets that can be used by state agencies to support Source Water Assessment Programs (SWAPs). We would like to include your state in this comprehensive data inventory.

A questionnaire has been developed by Technology Planning and Management Corporation (TPMC), the prime contractor on this project that will help identify sources of Chemical Occurrence Watershed data that can support source water assessments. The questionnaire will be available online at http://www.drinkingh2o.com/swap. You may complete the questionnaire at the above website with or without support from TPMC scientists. Alternatively, a matrix format has been developed with the survey questions. At your request, an electronic version of this matrix will be provided.

Within the next two weeks, you will be contacted by Ms. Kristyn Stevens (TPMC) to arrange for accessing the password-secured system and for completing the survey. After your survey has been completed, we may need to contact you if further clarification is necessary.

We hope you will agree to participate in this project and thank you in advance for your consideration. Please contact Dr. Jose Sobrinho, TPMC's manager for this project, if you have any questions or would like additional information. You may contact Dr. Sobrinho at (781)-544-3189 or via e-mail at jsobrinho@tpmc.com.

Sincerely,

Traci Case
AWWARF Manager/Project "Chemical Occurrence Data sets for Source Water Assessment"

Enclosure

Chemical Occurrence Data for Source Water Assessments

Background

The Safe Drinking Water Act (SDWA) established in 1974 is designed to protect the quality of drinking water in the United States. Section 1453 of the 1996 SDWA Amendments requires all states to establish Source Water Assessment Programs (SWAP). Each state is required to have completed Source Water Assessments by 2003. This includes the submission of a plan to the United States Environmental Protection Agency (USEPA) detailing how they will delineate source water assessment areas, inventory significant contaminants in these areas, and determine the susceptibility of each public water supply to contamination. Currently all states have approved SWAP plans in place.

Source water assessments allow states to systematically address issues of potential contamination of public water supplies using existing data from established environmental programs. The chemical occurrence data developed and compiled for source water assessments is a strategic resource to accomplish this goal. Although an enormous amount of chemical data and information are currently collected, stored and managed elsewhere, data are not easily retrieved when needed for assessing source waters. As a result, AwwaRF is interested in creating a SWAP inventory including information on national databases, as well as data sets available through regional, state, and local programs and agencies. The information gathered from these sources would identify, list and characterize chemical occurrence data sets covering both surface and ground water. A well-characterized data set would require different levels of metadata including the following:

- Regions, watersheds, water bodies or aquifers
- Description of data
- Sampling methods
- Location and description of data sampling points
- Chemical groups. This may include nutrients, pesticides and semi-volatile organics compounds, volatile organics compounds, general organics (e.g. Biological Oxygen Demand (BOD), algal toxins), other inorganics, metals, radionuclides, particulates (e.g., sediment and turbidity), and general physiochemical (e.g., pH, perclorate)
- Analytical methods and associated detection limits
- Data set time frame
- Program title, objectives, and contact information
- Quality assurance and quality control procedures for sample collection and analysis
- Overall assessment of the quality of the data

We designed and developed a survey that is posted on the project website, which is designed to facilitate the input of the above information. This will include descriptive information as well as a series of "yes/no" questions such as "Is information on analytical methods/methods detection limits available?" Figure B.1 provides an example of the survey questions.

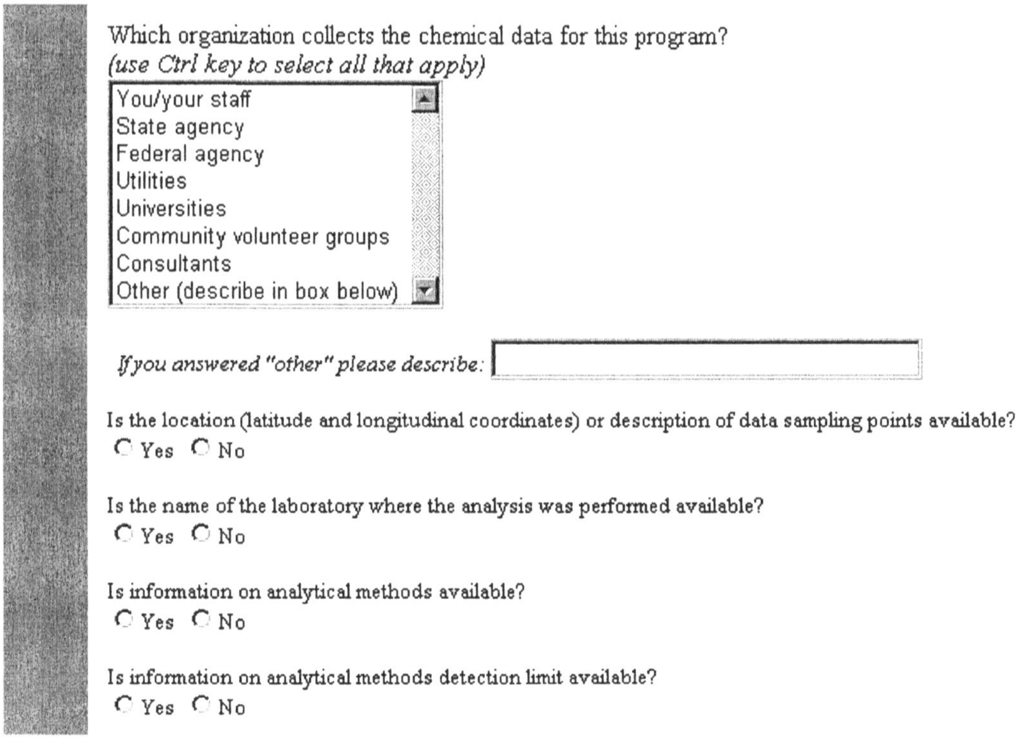

Which organization collects the chemical data for this program?
(use Ctrl key to select all that apply)

You/your staff
State agency
Federal agency
Utilities
Universities
Community volunteer groups
Consultants
Other (describe in box below)

If you answered "other" please describe:

Is the location (latitude and longitudinal coordinates) or description of data sampling points available?
Yes No

Is the name of the laboratory where the analysis was performed available?
Yes No

Is information on analytical methods available?
Yes No

Is information on analytical methods detection limit available?
Yes No

Figure B.1 Example of the survey questions

A web module with website links to online databases, agencies, data set summaries, and regulatory information will be constructed as the final product for this research project. During the project, this website will be accessed exclusively by the Project Team, AwwaRF and states participating in the research.

The development and implementation of this inventory of chemical data for SWAPs will have numerous practical benefits to the drinking water community. This inventory will help states to perform source water assessments, determine potential contaminants and perform susceptibility analyses. In addition, this tool may also assist agencies in improving regulations and programs, facilitate future research programs, and promote public awareness of the relationship between human activities and the protection of public water supplies.

APPENDIX C
STATE SURVEY

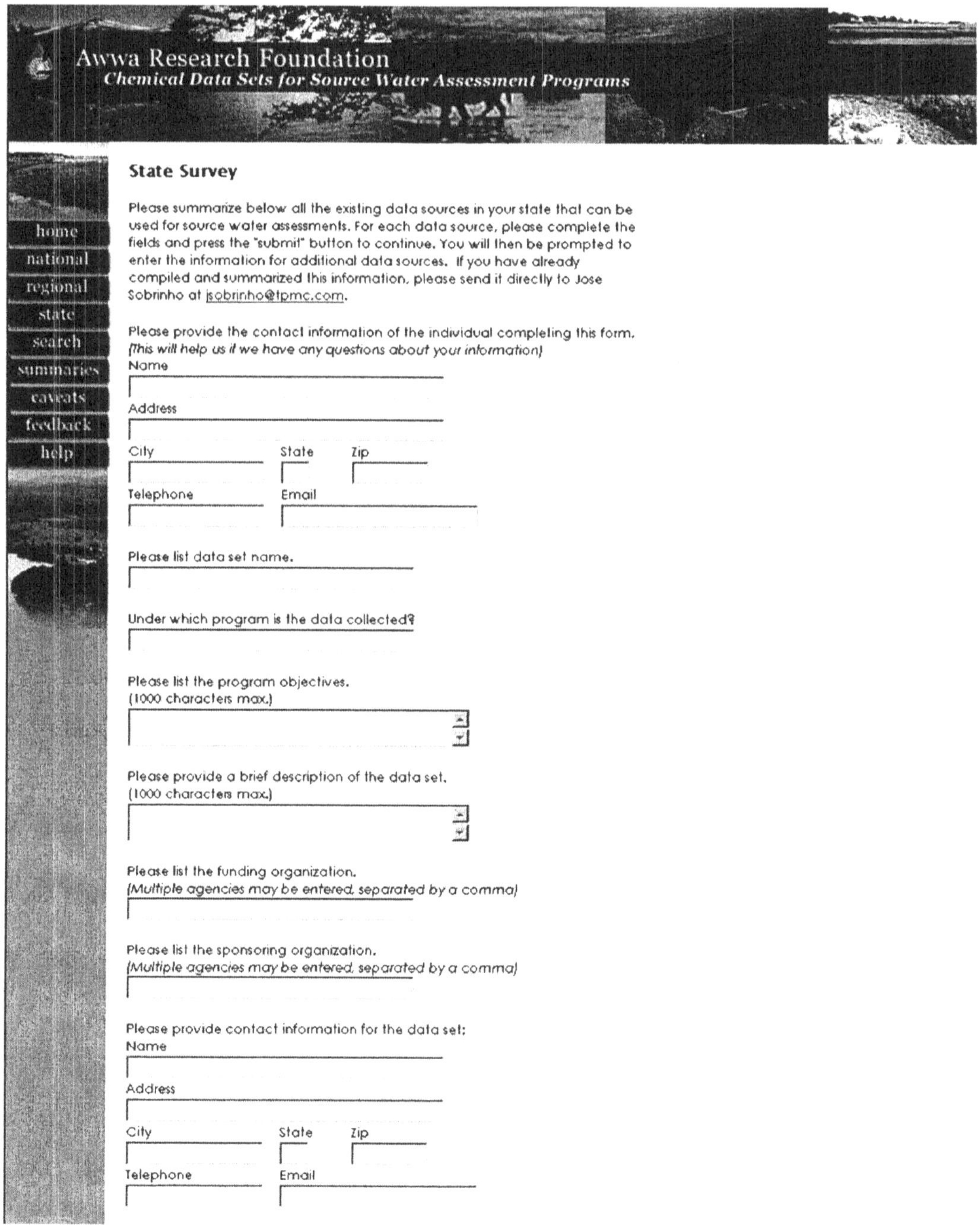

Awwa Research Foundation

Chemical Data Sets for Source Water Assessment Programs

home
national
regional
state
search
summaries
caveats
feedback
help

State Survey

Please summarize below all the existing data sources in your state that can be used for source water assessments. For each data source, please complete the fields and press the "submit" button to continue. You will then be prompted to enter the information for additional data sources. If you have already compiled and summarized this information, please send it directly to Jose Sobrinho at jsobrinho@tpmc.com.

Please provide the contact information of the individual completing this form.
(This will help us if we have any questions about your information)

Name

Address

City State Zip

Telephone Email

Please list data set name.

Under which program is the data collected?

Please list the program objectives.
(1000 characters max.)

Please provide a brief description of the data set.
(1000 characters max.)

Please list the funding organization.
(Multiple agencies may be entered, separated by a comma)

Please list the sponsoring organization.
(Multiple agencies may be entered, separated by a comma)

Please provide contact information for the data set:

Name

Address

City State Zip

Telephone Email

STATE SURVEY, continued

If the data set is available online, please provide the full URL.
(Example: http://www.drinkingh2o.com. If the data set is not available online, please leave this field blank.)

What is the geographic extent of the data set?
Aquifer

Which EPA Region(s) does the data include?
(use Ctrl key to select multiple entries)
Region 01
Region 02
Region 03
Region 04
Region 05

What state(s) are included in the data?
(use Ctrl key to select multiple entries)
Alabama
Alaska
Arizona
Arkansas
California

Does the data set include a description of the sampling area (e.g., Mid-Atlantic, Southeastern Massachusetts)?
(1000 characters maximum.)

What time frame is covered by the data set?
(Please provide both a start date and an end date. MM/DD/YYYY)
Start
End

Which of the following chemical groups are included in the dataset?
(Click on a group for more information. Note: specific parameters listed for each category are examples and are not meant to be comprehensive. There may be other parameters not listed that can be included in each category.)

- General physiochemical
- Nutrients
- Metals
- Other inorganics
- General organics
- Pesticides and SOCs
- VOCs
- Radionuclides
- Particulates

Are the following available for this data set?

Yes	No	Not sure	
○	○	○	Sampling methods
○	○	○	Location or description of data sampling points (e.g., latitude/longitude, LORAN)
○	○	○	Analytical methods
○	○	○	Detection limits
○	○	○	Quality assurance/Quality control procedures for sample collection
○	○	○	Quality assurance/Quality control procedures for analysis

Please provide an overall assessment of the quality of the data set.
(1000 characters maximum.)

Thank you for your assistance with this project. We look forward to sharing the results with you and hope that this will be useful to you when completing Source Water Assessments.

Submit | Reset

home | national | regional | state | search | summaries | feedback | help | state survey | site info

Updated 07/25/2002. Site development by Technology Planning & Management Corporation.

APPENDIX D
NATIONAL DATABASE SUMMARY

Storage and Retrieval (STORET) Legacy Data Center
Description: The database stores raw biological, chemical, and physical data on surface and ground water collected by federal, state and local agencies, Indian Tribes, volunteer groups, academics, and others.
Database URL: http://www.epa.gov/storet
Chemical groups: Nutrients, Volatile Organic Compounds (VOCs), other inorganics, radionuclides, general physiochemical, pesticides and Synthetic Organic Compounds (SOCs), general organics, metals, and particulates.
Period of performance: 01/01/1960 - 12/31/1998
Contact: http://www.epa.gov/storet/contacts_frame.html

Modernized Storage and Retrieval (STORET)
Description: The database stores raw biological, chemical, and physical data on surface and ground water collected by federal, state and local agencies, Indian Tribes, volunteer groups, academics, and others.
Database URL: http://www.epa.gov/storet
Chemical groups: Nutrients, VOCs, other inorganics, radionuclides, general physiochemical, pesticides and SOCs, general organics, metals, and particulates.
Period of performance: 01/01/1999 - Present
Contact: http://www.epa.gov/storet/contacts_frame.html

Safe Drinking Water Information System (SDWIS)
Description: This is an USEPA national database storing routine information about the nation's drinking water. It maintains state-level information about water systems, sample data, violations, and enforcement actions.
Database URL: http://www.epa.gov/enviro/html/sdwis/
Chemical groups: Nutrients, VOCs, other inorganics, radionuclides, general physiochemical, pesticides and SOCs, general organics, metals, and particulates.
Period of performance: 01/01/1993 - 01/01/2001
Contact: USEPA's Freedom Of Information Act (FOIA) Office
USEPA
401 M St. SW (1105)
Washington, DC, 20460
202-260-4048

National Water Information System (NWIS)
Description: The NWIS Web water-quality database is a compilation of several USGS databases. The database includes data on chemical concentrations in water, sediment, and aquatic organism tissues and related quality control data, biological data, and daily stream flow and temperature.
Database URL: http://water.usgs.gov/nwis
Chemical groups: Nutrients, VOCs, other inorganics, radionuclides, general physiochemical, pesticides and SOCs, general organics, metals, and particulates.
Period of performance: 01/01/1960 - present
Contact: h2oteam@usgs.gov

National Water Quality Assessment (NAWQA) Program
Description: The database contains the following data: chemical concentrations in water, sediment, and aquatic organism tissues and related quality control data; biological data on stream habitat and community data on fish, algae and invertebrates; site, well and network information (land use, soils, population density, etc); daily stream flow and temperature.
Database URL: http://water.usgs.gov/nawqa
Chemical groups: Nutrients, VOCs, other inorganics, radionuclides, pesticides and SOCs, and metals.
Period of performance: 01/01/1991 - present
Contact: Alex K. Williamson, NAWQA National Database Team Leader
US Geological Survey, Water Resources Div.
1201 Pacific # 600
Tacoma, WA, 98402
253-428-3600x2683
akwill@usgs.gov

Nutrient Criteria
Description: Contains ambient data from USEPA's Legacy STORET, the USGS NASQAN, NAWQA, and other sources such as universities and states/tribes.
Database URL: http://www.epa.gov/ost/criteria/nutrient/database/
Chemical groups: Nutrients
Period of performance: Not available
Contact: OW-GENERAL@epa.gov

National Contaminant Occurrence Database (NCOD)
Description: Presents information on unregulated contaminants in public drinking water from USEPA's SDWIS and ambient source water from NWIS provided by USGS.
Database URL: http://www.epa.gov/ncod/
Chemical groups: Nutrients, VOCs, other inorganics, radionuclides, general physiochemical, pesticides and SOCs, general organics, metals, and particulates.
Period of performance: 01/01/1983 - 01/01/2001
Contact: http://www.epa.gov/ncod/ncod_feedback.html

National Stream Quality Accounting Network (NASQAN)
Description: USGS database containing water chemistry and sediment data collected from several major rivers throughout the USA.
Database URL: http://water.usgs.gov/nasqan/

Chemical groups: Nutrients, other inorganics, general physiochemical, pesticides and SOCs, general organics, metals, and particulates.
Period of performance: 01/01/1973 – Present
Contact: Rick Hooper
U.S. Geological Survey
508-490-5065
rhooper@usgs.gov

APPENDIX E
REGIONAL DATA SET SUMMARY

Kansas Stream Chemistry Database
Description: The Kansas Stream Chemistry Database contains some 750,000 data points describing water quality conditions at more than 500 stream locations in Kansas and bordering regions of adjoining states. The data is stored in DB2 format on an IBM AS-400 computer system maintained by the Kansas Department of Health and Environment.
States: Kansas, Missouri, Nebraska, and Oklahoma
Database URL: Not available
Chemical groups: Nutrients, other inorganics, radionuclides, general physiochemical, pesticides and SOCs, metals, and particulates.
Period of Performance: 07/05/1967-12/31/2001
Contact: Mike Butler or Anthony Stahl
Kansas Department of Health and Environment
1000 SW Jackson
Topeka, KS, 66612-1367
785-296-6603
mbutler@kdhe.state/ks.us or tstahl@kdhe.state.ks.us

Chesapeake Bay Program Water Quality Database
Description: The database includes measured and calculated physical, metal, other organics, general inorganics and particulates for 49 mainbay and approximately 150 tributary stations for Maryland, Virginia, and the District of Columbia.
States: Maryland and Virginia
Database URL: http://cobia.chesapeakebay.net/data/data_desc.cfm?DB=CBP_WQDB
Chemical groups: Nutrients, other inorganics, general physiochemical, general organics, and metals.
Period of Performance: 01/01/1984 - present
Contact: Ricky Bahner
Chesapeake Bay Program
410 Severn Ave., Suite 109
Annapolis, MD, 21403
800-9687-229 ext. 742
rbahner@chesapeakebay.net

Chesapeake Bay Toxics Program Database
Description: Stores pesticides, SOCs, VOCs, metals, other inorganics, and general organics data.
States: Maryland and Virginia
Database URL: http://cobia.chesapeakebay.net/data/data_desc.cfm?DB=CBP_TOXDB
Chemical groups: VOCs, other inorganics, pesticides and SOCs, general organics, and metals.
Period of Performance: 01/01/1973 - present

Contact: Kelly Shenk
Chesapeake Bay Program
410 Severn Ave., Suite 109
Annapolis, MD, 21403
800-9687-229 ext. 728
shenk.kelly@epa.gov

Alliance Citizen Monitoring Program Database
Description: Contains basic water quality and nutrient data.
States: Maryland, Pennsylvania, and Virginia
Database URL: http://www.acb-online.org/monitoring/site.cfm
Chemical groups: Nutrients and general physiochemical.
Period of Performance: 01/01/1985 - present
Contact: Alliance for the Chesapeake Bay
6600 York Road, Suite 100
Baltimore, MD, 21212
410-377-6270
acbmail@acb-online.org

Arkansas STORET
Description: This data has been developed over the past 30 years as part of the routine monitoring of the state's waterbodies.
States: Arkansas, Louisiana, Missouri, Oklahoma, and Texas
Database URL: Not available
Chemical groups: Nutrients, VOCs, other inorganics, general physiochemical, pesticides and SOCs, general organics, metals, and particulates.
Period of Performance: 01/01/70 - Present
Contact: Alan Price
Arkansas Department of Environmental Quality
8001 National Drive
Little Rock, AR, 72219
501-682-0744

New England Interstate Water Pollution Control Commission (NEIWPCC)
Description: Contains nutrient and general physiochemical data for lakes, rivers, reservoirs, ponds, and streams.
States: Connecticut, Massachusetts, Maine, New Hampshire, Rhode Island, and Vermont
Database URL: Not available
Chemical groups: Nutrients and general physiochemical.
Period of Performance: 01/01/1990-01/01/1999
Contact: Beth Card
New England Interstate Water Pollution Control Commission (NEIWPCC)
Boott Mills South, 100 Foot of John Street
Lowell, MA, 01852

(978) 323-7929
Bcard@neiwpcc.org

Occurrence and Distribution of Pesticides in Chesapeake Bay
Description: Compiles pesticide data measured in surface water, surface microlayer, groundwater, sediment, and biota of the Bay. Twelve pesticides are included in the database: alachlor, atrazine, carbofuran, chlordane, chlorpyrifos, chlorothanonil, diflubenzuron, linuron, metolachlor, permethrin, simzine, and toxaphene. Summary statistics with number of samples collected, maximum, minimum, and mean is provided for each station.
States: Maryland and Virginia
Database URL: http://www.agnic.org/cbp/index.html
Chemical groups: Pesticides and SOCs.
Period of Performance: 01/01/1976 - 01/01/1994
Contact: W. Edward Johnson
National Oceanic and Atmospheric Administration (NOAA)
1325 East West Hwy 8176 SSMC2
Silver Spring, MD, 20910-3283
301-713-0177
Ed.Johnson@noaa.gov

Groundwater Protection Council (GWPC) Environmental Information Management Suite Source Water Assessment Database Development*
Description: In 1992, the Groundwater Protection Council (GWPC) began developing the Environmental Information Management Suite (EIMS) to improve regulatory decision-making. This system is designed to assist states and industry in managing natural resources. The next major step in GWPC's EIMS program is the Source Water Assessment Database, currently utilized by Oklahoma, Nebraska and Nevada.
States: Oklahoma, Nebraska, and Nevada
Database URL: http://www.gwpc.org/EIMS/
Chemical groups: Not available
Period of Performance: Not available
Contact: Ground Water Protection Council
13208 N. MacArthur
Oklahoma City, OK 73142
405-516-4972
This data set is currently under development

Schuylkill and Delaware River Watershed Database
Description: Data was downloaded from the STORET, USGS NASQAN and USGS NAWQA websites as well as data compiled from local watershed organizations and water suppliers. Mainly this data is most robust for basic physical parameters, nutrients, and metals. Water supplier data provides a number of daily or weekly measurements for the past decade. Data spans the period of 1970 to 2002. Updates are made yearly to the database.
States: Delaware, New Jersey, New York, and Pennsylvania

Database URL: Not available
Chemical groups: Nutrients, VOCs, other inorganics, radionuclides, general physiochemical, pesticides and SOCs, general organics, metals, and particulates.
Period of Performance: 01/01/1970-01/01/2002
Contact: Christopher S. Crockett
Philadelphia Water Department
1101 Market St.
Philadelphia, PA, 19107
215-685-6234
chris.crockett@phila.gov

APPENDIX F
STATES DATA SET SUMMARY

ALABAMA

Reservoir Monitoring Data

Description: Physical, chemical, and biological data collected in accordance with the state's 5-year rotating river basin monitoring schedule. Reservoirs within the focus river basin are intensively sampled from April through October. Other reservoirs are sampled once or twice per year. Data includes water column profile data for Dissolved Oxygen (DO), pH, temperature, and conductivity. Composite water quality samples are collected from the photic zone and analyzed for Chlorophyll *a* (chl-a), nutrients, Total Suspended Solids (TSS), Total Dissolved Solids (TDS), turbidity, alkalinity, hardness, Total Organic Carbon (TOC), and fecal coliform.
Database URL: Not available
Chemical groups: Nutrients and general physiochemical.
Period of Performance: 04/01/1990 - 10/31/2001
Contact: Fred Leslie
Alabama Department of Environmental Management
P.O. Box 301463
Montgomery, AL, 36130-1463
334-260-2752
fal@adem.state.al.us

ALASKA

State of Alaska Public Water System Supervision (PWSS) Database

Description: Routine water quality monitoring for public water systems.
Database URL: Not available
Chemical groups: Nutrients, VOCs, other inorganics, radionuclides, pesticides and SOCs, general organics, metals, and particulates.
Period of Performance: 01/01/1980 - 12/31/2000
Contact: Maria Ridgeway
Alaska Department of Environmental Conservation/Drinking Water & Wastewater Program
555 Cordova Street
Anchorage, AK, 99501
907-269-7625
maria_ridgeway@envircon.state.ak.us

ARIZONA

Water Quality Division Database

Description: System permanently stores and maintains drinking water systems inventory, sample, and test data. The system also performs drinking water system violations and

compliance calculations according to federal and state regulations.
Database URL: Not available
Chemical groups: VOCs, other inorganics, general physiochemical, pesticides and SOCs, and particulates.
Period of Performance: Not available
Contact: Wayne Hood, Manager
ADEQ Drinking Water Section
3033 N. Central Ave., MO248A
Phoenix, AZ, 80512
602-207-4427
hood.wayne@ev.state.az.us

ARKANSAS
Arkansas STORET
Description: This database has been developed over the past 30 years as part of the routine monitoring of the state's waterbodies.
Database URL: Not available
Chemical groups: Nutrients, VOCs, other inorganics, general physiochemical, pesticides and SOCs, and particulates.
Period of Performance: 01/01/1970 - present
Contact: Alan Price
Arkansas Department of Environmental Quality
8001 National Drive
Little Rock, AR, 72219
501-682-0744
price@adeq.state.ar.us

CALIFORNIA
Malibu Creek Stream Team Database
Description: The database stores basic water quality and nutrients data collected at seven locations throughout the watershed.
Database URL: http://www.streamteam@healthebay.org/streamteamhome.asp
Chemical groups: Nutrients, general physiochemical, and particulates.
Period of Performance: 11/01/1998 - present
Contact: Malibu Creek Stream Team, Heal the Bay
3220 Nebraska Ave.
Santa Monica, CA, 90404
310-453-0395
streamteam@healthebay.org

Department of Water Resources Water Data Library
Description: Compiles water quality data for surface waters.

Database URL: http://wdl.water.ca.gov/default.asp
Chemical groups: Other inorganics, general physiochemical, pesticides and SOCs, general organics, and particulates.
Period of Performance: Not available
Contact: California Department of Water Resources
901 P Street
Sacramento, CA, 95814
916-651-9236

Department of Pesticides Regulation's Surface Water Monitoring Database
Description: Results of chemical analyses for pesticides in water samples taken from California rivers, creeks, urban streams, agricultural drains, the San Francisco Bay-Delta region, and urban stormwater runoff.
Database URL: http://www.cdpr.ca.gov/docs/surfwatr/surfdata.htm
Chemical groups: Pesticides and SOCs.
Period of Performance: 01/01/1997 - present
Contact: Keith Starner
California Department of Pesticide Regulation
1001 I Street, PO Box 4015
Sacramento, CA, 95812-4015
916-324-4167
kstarner@cdpr.ca.gov

COLORADO
Rivers of Colorado Water Watch Network Database
Description: Collection of basic water quality data, metals, nutrients, and several inorganics
Database URL: http://wildlife.state.co.us/riverwatch/Contents/new%20search_the_data.htm
Chemical groups: Nutrients, other inorganics, general physiochemical, and metals.
Period of Performance: 01/01/1990 - present
Contact: Ann Seiler
Colorado Department of Natural Resources
Colorado Division of Wildlife
1424 NE Frontage Rd.
Fort Collins, CO, 80524
970-472-4406
ann.seilen@state.co.us

CONNECTICUT
New England Interstate Water Pollution Control Commission (NEIWPCC)
Description: Contains nutrient and general physiochemical data for lakes, rivers, reservoirs, ponds, and streams.
Database URL: Not available

Chemical groups: Nutrients and general physiochemical.
Period of Performance: 01/01/1990 - 01/01/1999
Contact: Beth Card
New England Interstate Water Pollution Control Commission (NEIWPCC)
Boott Mills South
100 Foot of John Street
Lowell, MA, 01852
978-323-7929
Bcard@neiwpcc.org

DELAWARE

Delaware Public Water Systems Supply Database

Description: Selected MCL and SMCL chemical data generated by the Delaware Department of Health and Social Services - Office of Drinking Water.
Database URL: Not available
Chemical groups: Nutrients, VOCs, other inorganics, radionuclides, general physiochemical, pesticides and SOCs, general organics, and metals.
Period of Performance: (start date not available) - 01/28/2002
Contact: Edward Hallock
Delaware Department of Natural Resources (DNR) and Environmental Control
89 Kings Hwy
Dover, DE, 19901
302-739-5410
ehallock@state.de.us

Schuylkill and Delaware River Watershed Database

Description: Data was downloaded from the STORET, USGS NASQAN and USGS NAWQA websites, as well as data compiled from local watershed organizations and water suppliers. This data is most robust for basic physical parameters, nutrients, and metals. Water supplier data provides a number of daily or weekly measurements for the past decade. Data spans the period of 1970 to 2002. Updates are made yearly to the database.
Database URL: Not available
Chemical groups: Nutrients, VOCs, other inorganics, radionuclides, general physiochemical, pesticides and SOCs, general organics, metals, and particulates.
Period of Performance: 01/01/1970 - 01/01/2002
Contact: Christopher S. Crockett
Philadelphia Water Department
1101 Market St.
Philadelphia, PA, 19107
215-685-6234
chris.crockett@phila.gov

FLORIDA

Generalized Water Information System (GWIS3)

Description: Background groundwater quality data including major ions, nutrients, volatile organic compounds, pesticides, trace metals, base neutral, and basic water chemistry.

Database URL: http://www.dep.state.fl.us/water/monitoring/data.htm

Chemical groups: VOCs, other inorganics, general physiochemical, pesticides and SOCs, metals, and particulates.

Period of Performance: 01/01/1987 - present

Contact: Jay Silvanima
Florida Department of Environmental Protection
3900 Commonwealth Blvd, MS # 70
Tallahassee, FL, 32399-2400
850-921-9427
jay.silvanima@dep.state.fl.us

GEORGIA

Safe Drinking Water Information System/State

Description: SDWIS/State data system developed by states and the USEPA.

Database URL: Not available

Chemical groups: Nutrients, VOCs, other inorganics, general physiochemical, pesticides and SOCs, general organics, metals, and particulates.

Period of Performance: Not available

Contact: Doug Davenport
GA Department of Natural Resources
Environmental Protection Division
205 Jesse Hill Jr. Dr., SE, Suite 1362
Atlanta, GA, 30334
404-651-5162
doug_davenport@dnr.state.ga.us

Radold

Description: Radiological data for source and distributions system monitoring.

Database URL: Not available

Chemical groups: Radionuclides

Period of Performance: Not available

Contact: Brad Addison
GA Department of Natural Resources, Environmental Protection Division
205 Jesse Hill Jr. Dr., SE; Suite 1362
Atlanta, GA, 30334
404-651-5155
brad_addison@dnr.state.ga.us

HAWAII

Department of Health Water Quality Monitoring

Description: Based on required parameters to meet drinking water standards.

Database URL: Not available

Chemical groups: Nutrients, VOCs, other inorganics, radionuclides, pesticides and SOCs, and metals.

Period of Performance: 01/01/1974 - 04/30/2002

Contact: William Wong
Hawaii Department of Health
919 Ala Moana Blvd. # 308
Honolulu, HI, 96814
808-586-4258
wwong@eha.health.state.hi.us

IDAHO

Idaho Surface Water Quality Statewide Network

Description: Compilation of water quality for state streams.

Database URL: http://idaho.usgs.gov/public/wq/index.html

Chemical groups: Nutrients, other inorganics, general physiochemical, and particulates.

Period of Performance: 01/01/1990 - present

Contact: Ivalou O'Dell
USGS Water Resources of Idaho
230 Collins Road
Boise, ID, 83702
208-387-1325
ioodell@usgs.gov

ILLINOIS

Illinois Source Water Assessment Database

Description: Compiles information on source of drinking water (surface water, groundwater, or both), characteristics of the community water-supply wells and/or intakes, source and finished water quality, potential sources of contamination, susceptibility to potential contamination, and surface and groundwater protection efforts.

Database URL: http://idaho.usgs.gov/public/wq/index.html (site is not available to the public at the present time)

Chemical groups: Nutrients, VOCs, other inorganics, pesticides and SOCs, and metals.

Period of Performance: 01/01/1985 - present

Contact: Wade C. Boring
Illinois EPA
1021 North Grand Ave East
Springfield, IL, 62794-9276

217-785-4787
wade.boring@epa.state.il.us

The Lower Illinois River Basin NAWQA Study Unit
Description: Data on nutrients, selected inorganics, and trace elements
Database URL: http://il.water.usgs.gov/proj/lirb/pubs/pdfs/topicalbook.pdf
Chemical groups: Nutrients, other inorganics, and metals.
Period of Performance: 01/01/1984 - 01/01/1991
Contact: Kelly Warner
US Geological Survey
221 North Broadway Avenue
Urbana, IL, 61801
217-344-0037

Radium in Groundwater from Public Supply Aquifers in Northern Illinois
Description: Compilation of radium data collected from wells.
Database URL: http://il.water.usgs.gov/proj/gwstudies/radium/
Chemical groups: Radionuclides
Period of Performance: Not available
Contact: Bob Kay
US Geological Survey
1420 Sycamore Road
De Kalb, IL, 60115
815-756-9207
rtkay@usgs.gov

INDIANA
Drinking Water Compliance Database and SDWIS/State
Description: The data set incorporates the inventory information of each public water system. It also incorporates the violations and enforcement actions taken for each system. All chemical data submitted by each system are entered into the database.
Database URL: Not available
Chemical groups: Nutrients, VOCs, other inorganics, radionuclides, general physiochemical, pesticides and SOCs, general organics, metals, and particulates.
Period of Performance: 01/01/1993 - 12/31/2001
Contact: Wayne Wang
Indiana Department of Environmental Management
100 N. Senate Ave.
P.O. Box 7148
Indianapolis, IN, 46207-7148
317-308-3296
wwang@dem.state.in.us

IOWA

Iowa STORET

Description: The database incorporates physical, biological, and habitat data

Database URL: http://wqm.igsb.uiowa.edu/storet/

Chemical groups: Nutrients, VOCs, other inorganics, general physiochemical, pesticides and SOCs, metals, and particulates.

Period of Performance: 10/01/1999 - present

Contact: Mary Skopec
Iowa Department of Natural Resources
Energy & Geological Resources Division, Water
Monitoring Section
109 Trowbridge Hall
Iowa City, Iowa, IA, 52242-1319
319-335-1575
mskopec@igsb.uiowa.edu

Environmental Monitoring for Public Access and Community Tracking/Des Moines Database

Description: Selected chemical and biological data

Database URL: http://www.dmww.com/empact_p2.asp

Chemical groups: Nutrients, other inorganics, general physiochemical, pesticides and SOCs, metals, and particulates.

Period of Performance: 06/01/1999 - present

Contact: Des Moines Water Works
2201 Valley Drive
Des Moines, IA, 50321-1190
515-283-8700
information@dmww.com

KANSAS

Kansas Stream Chemistry Database Description: The Kansas Stream Chemistry Database contains some 750,000 data points describing water quality conditions at more than 500 stream locations in Kansas and bordering regions of adjoining states. The data is stored in DB2 format on an IBM AS-400 computer system maintained by the Kansas Department of Health and Environment.

Database URL: Not available

Chemical groups: Nutrients, other inorganics, radionuclides, general physiochemical, pesticides and SOCs, metals, and particulates.

Period of Performance: 07/05/1967 - 12/31/2001

Contact: Mike Butler or Anthony Stahl
Kansas Department of Health and Environment

1000 SW Jackson
Topeka, KS, 66612-1367
785-296-6603
mbutler@kdhe.state/ks.us or tstahl@kdhe.state.ks.us

Kansas Lake and Wetland Water Quality Database
Description: The Kansas Lake and Wetland Water Quality Database contains some 150,000 records describing water quality conditions in approximately 430 lakes and wetlands in Kansas. These records are stored in DB2 format on an IBM AS-400 computer system maintained by the Kansas Department of Health and Environment.
Database URL: Not available
Chemical groups: Nutrients, other inorganics, general physiochemical, pesticides and SOCs, metals, particulates
Period of Performance: 05/01/1975 - 12/31/2001
Contact: Ed Carney
Kansas Department of Health and Environment
1000 SW Jackson
Topeka, KS, 66612-1367
785-296-6603
ecarney@kdhe.state.ks.us

Kansas Groundwater Quality Database
Description: The Kansas Groundwater Quality Database contains more than 200,000 data points describing groundwater quality conditions at approximately 300 locations in Kansas. The data is stored in DB2 format on an IBM AS-400 computer system maintained by the Kansas Department of Health and Environment.
Database URL: Not available
Chemical groups: Nutrients, VOCs, other inorganics, radionuclides, general physiochemical, pesticides and SOCs, and metals.
Period of Performance: 04/10/1985 - 12/31/2001
Contact: Bryan Bain
Kansas Department of Health and Environment
1000 SW Jackson
Topeka, KS, 66612-1367
785-296-6603
bbain@kdhe.state.ks.us

KENTUCKY
Compass
Description: All Department of Environmental Protection (DEP) environmental data -- chemistry, biological, habitat, toxicity, quantity, etc.
Database URL: Not available

Chemical groups: Nutrients, VOCs, other inorganics, general physiochemical, pesticides and SOCs, general organics, metals, and particulates.
Period of Performance: 01/01/1990 - 06/30/2002
Contact: Rob Thorne
Kentucky Department for Environmental Protection
14 Reilly Road
Frankfort, KY, 40601
502-564-6716
robert.thorne@mail.state.ky.us

The Kentucky Lake Long-Term Monitoring Program Data set
Description: It stores basic water quality data collected from 12 sampling stations and several embayments of the lake.
Database URL: http://www.murraystate.edu/qacd/cos/crr/crr-long.htm
Chemical groups: Nutrients, other inorganics, general physiochemical, and particulates.
Period of Performance: 07/01/1988 - present
Contact: Dr. David S. White, Program Coordinator
Center for Reservoir Research, Hancock Biological Station
561 Emma Drive
Murray, KY, 42071
270-474-2272
david.white@murraystate.edu

The Lake Barkley Database
Description: Water quality data collected from 12 sampling stations.
Database URL: http://www.murraystate.edu/qacd/cos/crr/crr-long.htm
Chemical groups: Nutrients, other inorganics, general physiochemical, and particulates.
Period of Performance: 07/01/1988 - present
Contact: Dr. David S. White, Program Coordinator
Center for Reservoir Research, Hancock Biological Station
561 Emma Drive
Murray, KY, 42071
270-474-2272
david.white@murraystate.edu

Ledbetter and Panther Watersheds Database
Description: It stores basic water quality data from several stations located in the two watersheds.
Database URL: http://www.murraystate.edu/qacd/cos/crr/crr-long.htm
Chemical groups: Nutrients, other inorganics, general physiochemical, and particulates.
Period of Performance: 01/01/1995 - present
Contact: Dr. David S. White, Program Coordinator

Center for Reservoir Research, Hancock Biological Station
561 Emma Drive
Murray, KY, 42071
270-474-2272
david.white@murraystate.edu

LOUISIANA
Arkansas STORET
Description: This data has been developed over the past 30 years as part of the routine monitoring of the state's waterbodies.
Database URL: Not available
Chemical groups: Nutrients, VOCs, other inorganics, and general physiochemical.
Period of Performance: 01/01/1970 – present
Contact: Alan Price
Arkansas Department of Environmental Quality
8001 National Drive
Little Rock, AR, 72219
501-682-0744
price@adeq.state.ar.us

MAINE
Public Water Supply Water Quality Data
Description: Water quality data from initial source approval samples (fairly comprehensive testing) followed by water system samples, primarily for acute contaminants, over time. Some periodic sampling of the system for chronic contaminants is included, depending on waiver status.
Database URL: Not available
Chemical groups: Nutrients, VOCs, other inorganics, radionuclides, general physiochemical, pesticides and SOCs, general organics, metals, and particulates.
Period of Performance: 01/01/1980 - 01/01/2002
Contact: Bob Peterson
Maine Drinking Water Program, 10 SHS
Augusta, ME, 04333
207-387-1979
Bob.Peterson@state.me.us

Public Educational Access to Environmental Information in Maine (PEARL)
Description: The data includes time series data on nitrogen and phosphorus concentrations, as well as lake transparency (Secci disk) and sometimes temperature readings. Data set is maintained at the University of Maine, and is accessible through a mapping web site. http://pearl.spatial.maine.edu/
Database URL: http://pearl.spatial.maine.edu/

Chemical groups: Nutrients and general physiochemical.
Period of Performance: 01/01/1980 - 12/31/2001
Contact: Tom Noonan
University of Maine, Dept of Spatial Engineering
Orono, ME, 04469
tnoonan@spatial.maine.edu

New England Interstate Water Pollution Control Commission (NEIWPCC)
Description: Contains nutrient and general physiochemical data for lakes, rivers, reservoirs, ponds, and streams.
Database URL: Not available
Chemical groups: Nutrients, general physiochemical
Period of Performance: 01/01/1990 - 01/01/1999
Contact: Beth Card
New England Interstate Water Pollution Control Commission (NEIWPCC)
Boott Mills South
100 Foot of John Street
Lowell, MA, 01852
978-323-7929
Bcard@neiwpcc.org

MARYLAND
Drinking Water Monitoring Data - SDWIS
Description: Regulated Inorganic Chemicals (IOC), VOC, and SOC data including Trihalomethanes (THMs).
Database URL: Not available
Chemical groups: Nutrients, VOCs, other inorganics, radionuclides, general physiochemical, pesticides and SOCs, general organics, metals, and particulates.
Period of Performance: 01/01/1993 - 01/01/2001
Contact: Maryland Department of the Environment
2500 Broening Highway
Baltimore, MD, 21224
410-631-3000

Maryland Department of the Environment (MDE)/Department of Natural Resources (DNR) Core Trend Data
Description: Samples include nutrient data, physical characteristics, suspended sediment, and some bacteriological data.
Database URL: Not available.
Chemical groups: Nutrients, other inorganics, general physiochemical, and particulates.
Period of Performance: 01/01/1974 - 01/01/2001
Contact: Maryland Department of Natural Resources

580 Taylor Avenue
Annapolis, MD, 21401
410-260-8100

Chesapeake Bay Program Water Quality Database
Description: The database includes measured and calculated physical, metal, other organics, general inorganics and particulates for 49 main bay and approximately 150 tributary stations for Maryland, Virginia, and the District of Columbia.
Database URL: http://cobia.chesapeakebay.net/data/data_desc.cfm?DB=CBP_WQDB
Chemical groups: Nutrients, other inorganics, general physiochemical, general organics, and metals.
Period of Performance: 01/01/1984 - present
Contact: Ricky Bahner
Chesapeake Bay Program
410 Severn Av., Suite 109
Annapolis, MD, 21403
800-9687-229 ext. 742
rbahner@chesapeakebay.net

Chesapeake Bay Toxics Program Database
Description: The database stores pesticides, SOCs, VOCs, metals, other inorganics, and general organics data.
Database URL: http://cobia.chesapeakebay.net/data/data_desc.cfm?DB=CBP_TOXDB
Chemical groups: VOCs, other inorganics, pesticides and SOCs, general organics, and metals.
Period of Performance: 01/01/1973 - present
Contact: Kelly Shenk
Chesapeake Bay Program
410 Severn Av., Suite 109
Annapolis, MD, 21403
800-9687-229 ext. 728
shenk.kelly@epa.gov

Alliance Citizen Monitoring Program Database
Description: Contains basic water quality and nutrients data.
Database URL: http://www.acb-online.org/monitoring/site.cfm
Chemical groups: Nutrients and general physiochemical.
Period of Performance: 01/01/1985 - present
Contact: Alliance for the Chesapeake Bay
6600 York Road, Suite 100
Baltimore, MD, 21212
410-377-6270
acbmail@acb-online.org

Occurrence and Distribution of Pesticides in Chesapeake Bay
Description: Compiles pesticide data measured in surface water, surface microlayer, groundwater, sediment, and biota of the Bay. Twelve pesticides are included in the database: alachlor, atrazine, carbofuran, chlordane, chlorpyrifos, chlorothanonil, diflubenzuron, linuron, metolachlor, permethrin, simzine, and toxaphene. Summary statistics with number of samples collected, maximum, minimum, and mean is provided for each station.
Database URL: http://www.agnic.org/cbp/index.html
Chemical groups: Pesticides and SOCs.
Period of Performance: 01/01/1976 - 01/01/1994
Contact: W. Edward Johnson
NOAA, 1325 East West Hwy 8176 SSMC2
Silver Spring, MD, 20910-3283
301-713-0177
Ed.Johnson@noaa.gov

MASSACHUSETTS
Water Quality Testing System
Description: Public Water System Inventory, Violations, DEP Enforcements, Compliance Monitoring data.
Database URL: Not available
Chemical groups: Nutrients, VOCs, other inorganics, radionuclides, pesticides and SOCs, metals
Period of Performance: 04/01/1993 - 02/11/2002
Contact: Damon Guterman
DEP, 1 Winter Street, 6th Floor
Boston, MA, 02108
617-574-6811
Damon.Guterman@state.ma.us

New England Interstate Water Pollution Control Commission (NEIWPCC)
Description: Contains nutrient and general physiochemical data for lakes, rivers, reservoirs, ponds, and streams.
Database URL: Not available
Chemical groups: Nutrients and general physiochemical.
Period of Performance: 01/01/1990 - 01/01/1999
Contact: Beth Card
New England Interstate Water Pollution Control Commission (NEIWPCC)
Boott Mills South
100 Foot of John Street
Lowell, MA, 01852
978-323-7929
Bcard@neiwpcc.org

Site/Reportable Releases Look Up
Description: Conduct an online search of sites in the Commonwealth where a release of oil or hazardous material has been reported to the Department of Environmental Protection (DEP). This database will be useful for people who want to see what sites DEP has listed for their city or town. Download DEP's databases of sites and spills. These databases provide more extensive information than that available in the online search and allow you to design specific queries in your database or spreadsheet program.
Database URL: http://www.state.ma.us/dep/bwsc/sites/report.htm or
http://www.state.ma.us/dep/sites/sdown.htm
Chemical groups: VOCs, other inorganics, pesticides and SOCs, general organics, and metals.
Period of Performance: Not available
Contact: Bureau of Waste Site Cleanup / MA DEP
One Winter Street - 7th Floor
Boston, MA, 02108
617-292-5500
bwsc.information@state.ma.us

Underground Storage Tank (UST) Registry
Description: The Department of Fire Services, Office of the State Fire Marshal administers the Underground Storage Tank Compliance program through its Technical Services Division. Code Compliance officers are responsible for enforcing UST regulations and overseeing the installation, operation, removal and disposal of USTs. The Office of the State Fire Marshal provides technical assistance to the public and Fire Service officials who assist in the administration of the UST Compliance program.
Database URL: http://db.state.ma.us/Dfs/ustQueryPage.asp
Chemical groups: VOCs and general organics.
Period of Performance: Not available
Contact: George Nice
Department of Fire Services
PO Box 1025, State Road
Stow, MA, 01775
978-567-3100
George.Nice@state.ma.us

MICHIGAN

Act 452 Part 201 Environmental Remediation Sites
Description: Contains statewide data set of sites that meet environmental contamination/ remediation criteria. Basic information about each site is included (name, location, type of facility, contaminants detected), when reported.
Database URL: http://www.deq.state.mi.us/erd1/index.jsp
Chemical groups: Nutrients, VOCs, other inorganics, radionuclides, general physiochemical, pesticides and SOCs, general organics, metals, and particulates.

Period of Performance: 06/01/1990 - 12/15/2001
Contact: Soheil Ghavami
MDEQ, Environmental Response Division
PO Box 30426
Lansing, MI, 48909-7926
517-373-9541
ghavamis@michigan.gov

Leaking Underground Storage Tanks (LUST)
Description: Because all LUSTs are to be registered on the Storage Tank lists prior, the LUST list contains a facility ID to link to the master list, whether the site remains open or is closed, date of release, address of release location.
Database URL: http://www.michigan.gov/deq/1,1607,7-135-3306_3334_3570---,00.html
Chemical groups: VOCs
Period of Performance: 01/01/1980 - 12/31/2001
Contact: Roger Przbysz
Michigan DEQ, Storage Tank Division
Lansing, MI, 48909-7657
517-373-8168

National Pollutant Discharge Elimination System (NPDES) Permit List
Description: Inventory of each facility/entity that wishes to discharge to surface water.
Database URL: http://www.deq.state.mi.us/swq/permits/active/active/htm
Chemical groups: Nutrients, VOCs, other inorganics, radionuclides, general physiochemical, pesticides and SOCs, general organics, metals, and particulates.
Period of Performance: 03/29/1974 - 12/31/2001
Contact: Fred Cowles
Michigan Department of Environmental Quality
525 W. Allegan
Lansing, MI, 48913
517-373-4633
cowlesf@michigan.gov

WaterChem Database
Description: The WaterChem database was created by the Michigan Department of Environmental Quality (MDEQ) Drinking Water and Radiological Protection Division (DWRP) to fill a need at the state level for a repository to host drinking water chemistry and microbiological results. The water sample results, of which the database is composed, are all from potable drinking water supplies. These sources range in size from single-family homes, to schools and business, up to major cities.
Database URL: Not available
Chemical groups: Nutrients, VOCs, other inorganics, radionuclides, general physiochemical,

pesticides and SOCs, general organics, metals, and particulates.
Period of Performance: 01/01/1993 - present
Contact: Kristen Philip
MDEQ-DWRP
525 W. Allegan
Lansing, MI, 48909-8130
517-241-1238
philipk@michigan.gov

MINNESOTA

Public Water Supply Supervision Program Compliance Monitoring

Description: Contains the results of compliance monitoring required by the federal regulations relating to the federal Safe Drinking Water Act. Most of the data relate to the quality of water entering a public water supply distribution system and are not always reflective of raw water quality obtained from a specific source of drinking water. However, a subset of these data reflect source water and are being used to provide a historical perspective of water quality. A summary of source water quality is included in source water assessments.
Database URL: Not available
Chemical groups: Nutrients, VOCs, radionuclides, pesticides and SOCs, metals, and particulates.
Period of Performance: 01/01/1996 - present
Contact: Doug Mandy
Minnesota Department of Health
121 East Seventh Place
Saint Paul, MN, 55101
651-215-0757
douglas.mandy@health.state.mn.us

Nitrate Nitrogen Analyses from Private Water Supply Wells

Description: Nitrate nitrogen analyses from private water supply wells. Data are collected by the Water Well Construction program at the Minnesota Department of Health (MDH) and by county health agencies. A limited amount of data come from other state agency programs.
Database URL: Not available
Chemical groups: Nutrients
Period of Performance: 01/01/1990 - present
Contact: Sheila Grow
Minnesota Department of Health
121 East Seventh Place
Saint Paul, MN, 55101
651-215-0768

Groundwater Monitoring and Assessment Program (GWMAP) Database
Description: Compilation of data on metals, organic compounds, and major cations and anions
Database URL: http://www.pca.state.mn.us/water/groundwater/gwmap/gwbaseline.html
Chemical groups: VOCs, other inorganics, general physiochemical, general organics, and metals.
Period of Performance: 01/01/1991 - 01/01/1996
Contact: Mike Trojan
Minnesota Pollution Control Agency, 520 Lafayette Road
St. Paul, MN, 55155-4194
651-297-5219
mike.trojan@pca.state.mn.us

Statewide Surface Water Monitoring for Nutrients and Pesticides
Description: Compiles pesticides and nutrients data.
Database URL: http://www.mda.state.mn.us/appd/ace/maace.htm
Chemical groups: Nutrients, pesticides and SOCs.
Period of Performance: 01/01/1992 - present
Contact: John Hines
Minnesota Department of Agriculture
90 West Plato Boulevard
St. Paul, MN, 55107
651-297-3994
John W.Hines@state.mn.us

MISSISSIPPI
Source Water Assessment Program (SWAP) Geographic Information System (GIS)
Description: The data is stored in ARC/INFO, which allows for various GIS applications and the generation of reports and maps. Efforts will commence on April 15, 2002 to design a new relational SWAP database. The new database will have analytical data querying capabilities; the present SWAP database does not have this capability.
Database URL: Not available
Chemical groups: Not available
Period of Performance: Not available
Contact: James Smith
Mississippi Department of Environmental Quality
P.O. Box 10385
Jackson, MS, 39289
601-961-5641
James_Smith@deq.state.ms.us

MISSOURI

Kansas Stream Chemistry Database

Description: The Kansas Stream Chemistry Database contains some 750,000 data points describing water quality conditions at more than 500 stream locations in Kansas and bordering regions of adjoining states. The data is stored in DB2 format on an IBM AS-400 computer system maintained by the Kansas Department of Health and Environment.

Database URL: Not available

Chemical groups: Nutrients, other inorganics, radionuclides, general physiochemical, pesticides and SOCs, metals, and particulates.

Period of Performance: 07/05/1967 - 12/31/2001

Contact: Mike Butler or Anthony Stahl
Kansas Department of Health and Environment
1000 SW Jackson
Topeka, KS, 66612-1367
785-296-6603
mbutler@kdhe.state/ks.us or tstahl@kdhe.state.ks.us

Arkansas STORET

Description: This data has been developed over the past 30 years as part of the routine monitoring of the state's waterbodies.

Database URL: Not available

Chemical groups: Nutrients, VOCs, other inorganics, general physiochemical, pesticides and SOCs, general organics, metals, and particulates.

Period of Performance: 01/01/1970 - present

Contact: Alan Price
Arkansas Department of Environmental Quality
8001 National Drive
Little Rock, AR, 72219
501-682-0744
price@adeq.state.ar.us

Missouri Stream Team Database

Description: Compiles basic water quality including general physiochemical parameters, nutrients, and few metals.

Database URL: http://www.mostreamteam.org/vmsearch.html

Period of Performance: 01/01/1993 - present

Contact: Sherry Fischer
Stream Team Unit, Missouri Department of Conservation
PO Box 180
Jefferson City, MO, 65102-0180
573-751-4115, ext 169
fischs@email.conservation.state.mo.us

MONTANA

Montana Natural Resource Information System (NRIS)

Description: This supplies access to the Water Information System, which stores groundwater data. Recently this application was replaced by data available through the thematic NRIS Data Mapper. This tool provides primary access to data used by the Montana Source Water Assessment Program (and others).

Database URL: http://nris.state.mt.us/mapper/

Chemical groups: Nutrients, general physiochemical, and metals.

Period of Performance: Not available

Contact: Joe Meek
SWAP Section, Pollution Prevention Bureau
MT Dept. Of Env. Quality
Metcalf Building, Box 200901
Helena, MT, 59620
(406) 444-4806
jmeek@state.mt.us

NEBRASKA

Kansas Stream Chemistry Database

Description: The Kansas Stream Chemistry Database contains some 750,000 data points describing water quality conditions at more than 500 stream locations in Kansas and bordering regions of adjoining states. The data is stored in DB2 format on an IBM AS-400 computer system maintained by the Kansas Department of Health and Environment.

Database URL: Not available

Chemical groups: Nutrients, other inorganics, radionuclides, general physiochemical, pesticides and SOCs, metals, and particulates.

Period of Performance: 07/05/1967 - 12/31/2001

Contact: Mike Butler or Anthony Stahl
Kansas Department of Health and Environment
1000 SW Jackson
Topeka, KS, 66612-1367
785-296-6603
mbutler@kdhe.state/ks.us or tstahl@kdhe.state.ks.us

Quality-Assessed Agrichemical Contaminant Database for Nebraska Groundwater

Description: Contains groundwater nitrate and pesticide data that have been compiled from federal, state, and local agencies as well as the University of Nebraska.

Database URL: http://nrcnt3.dnr.state.ne.us/clearinghouse/index.asp

Chemical groups: Pesticides and SOCs.

Period of Performance: 01/01/1974 - 01/01/2002

Contact: Dottie Harrell

103 Natural Resources Hall
University of Nebraska-Lincoln
Lincoln, NE, 68583-0844
(402) 472-8213
dharrell2@unl.edu

NEVADA

State of Nevada Surface Water Monitoring Network - Streams

Description: Contains surface (streams) water quality data including metals, nutrients, basic chemistry, and particulates.
Database URL: http://ndep.state.nv.us/bwqp/monitoringmap.htm
Chemical groups: Nutrients, other inorganics, general physiochemical, metals, and particulates.
Period of Performance: 01/01/1990 - present
Contact: Glen Gentry, Branch Supervisor
Nevada Division of Environmental Protection, Bureau of Water Quality Planning
Water Quality Monitoring Branch
333 West Nye Lane, Suite 138
Carson City, NV, 89706-0851
775-687-4670
ggentry@ndep.state.nv.us

State of Nevada Surface Water Monitoring Network – Lakes and Reservoirs

Description: Contains chlorophyll, metals and nutrients data for lakes and reservoirs.
Database URL: http://ndep.state.nv.us/bwqp/lakemap.htm
Chemical groups: Nutrients and metals.
Period of Performance: 01/01/1990 - present
Contact: Glen Gentry, Branch Supervisor
Nevada Division of Environmental Protection, and Bureau of Water Quality Planning
Water Quality Monitoring Branch
333 West Nye Lane, Suite 138
Carson City, NV, 89706-0851
775-687-4670
ggentry@ndep.state.nv.us

NEW HAMPSHIRE

Watershed Surface Water Quality Data

Description: Currently we have approximately 40 data sets that pertain to chemical and biological sampling of surface water. We are in the process of creating a database to consolidate these data sets and add appropriate metadata. Some of the data has been collected by volunteers under our volunteer sampling programs.
Database URL: Not available
Chemical groups: Nutrients, other inorganics, general physiochemical, general organics, metals,

and particulates.
Period of Performance: Not available
Contact: Deb Soule
NH Dept. of Env. Services
6 Hazen Drive, P.O. Box 95
Concord, NH, 03302-0095
603-271-8863
dsoule@des.state.nh.us

New England Interstate Water Pollution Control Commission (NEIWPCC)
Description: Contains nutrient and general physiochemical data for lakes, rivers, reservoirs, ponds, and streams.
Database URL: Not available
Chemical groups: Nutrients and general physiochemical.
Period of Performance: 01/01/1990 - 01/01/1999
Contact: Beth Card
New England Interstate Water Pollution Control Commission (NEIWPCC)
Boott Mills South
100 Foot of John Street
Lowell, MA, 01852
978-323-7929
Bcard@neiwpcc.org

NEW JERSEY
Schuylkill and Delaware River Watershed Database
Description: Data was downloaded from the STORET, USGS NASQAN and USGS NAWQA websites, as well as data compiled from local watershed organizations and water suppliers. This data is most robust for basic physical parameters, nutrients, and metals. Water supplier data provides a number of daily or weekly measurements for the past decade. Data spans the period of 1970 to 2002. Updates are made yearly to the database.
Database URL: Not available
Chemical groups: Nutrients, VOCs, other inorganics, radionuclides, general physiochemical, pesticides and SOCs, general organics, metals, and particulates.
Period of Performance: 01/01/1970 - 01/01/2002
Contact: Christopher S. Crockett
Philadelphia Water Department
1101 Market St.
Philadelphia, PA, 19107
215-685-6234
chris.crockett@phila.gov

New Jersey United States Geological Survey
Description: Compilation of water quality for state streams.
Database URL: http://nj.usgs.gov/gw/maps/index.html
Chemical groups: VOCs, radionuclides, pesticides and SOCs.
Period of Performance: Not available
Contact: Eric Vovinkel
U.S. Geological Survey, New Jersey District
810 Bear Tavern Road, Suite 206
West Trenton, NJ, 08628
vowinkel@usgs.gov

NEW YORK
Schuylkill and Delaware River Watershed Database
Description: Data was downloaded from the STORET, USGS NASQAN and USGS NAWQA websites, as well as data compiled from local watershed organizations and water suppliers. This data is most robust for basic physical parameters, nutrients, and metals. Water supplier data provides a number of daily or weekly measurements for the past decade. Data spans the period of 1970 to 2002. Updates are made yearly to the database.
Database URL: Not available
Chemical groups: Nutrients, VOCs, other inorganics, radionuclides, general physiochemical, pesticides and SOCs, general organics, metals, and particulates.
Period of Performance: 01/01/1970 - 01/01/2002
Contact: Christopher S. Crockett
Philadelphia Water Department
1101 Market St.
Philadelphia, PA, 19107
215-685-6234
chris.crockett@phila.gov

NORTH DAKOTA
Water Resources Data – Ground/Surface Water Data
Description: Ground/Surface Water Data provides access to well data that is available in North Dakota. This includes site information, lithologic, water level, and water chemistry data.
Database URL: http://www.swc.state.nd.us/dataresources.html
Chemical groups: Other inorganics and general physiochemical.
Period of Performance: 01/01/1970 - present
Contact: Chris Bader
North Dakota State Water Commission
900 East Boulevard Ave
Bismarck, ND, 58505-0850
701-328-2754
cbader@swc.state.nd.us

OHIO

Miami Conservancy District – Groundwater Quality Data

Description: Provides access to groundwater resources of the Great Miami River watershed (wells typically used for municipal drinking water or residential use). Also provides limited data from Ohio EPA, and USGS.

Database URL: http://www.miamiconservancy.org/GW2000/watermain.htm

Chemical groups: VOCs, other inorganics, general physiochemical, pesticides and SOCs.

Period of Performance: Not available

Contact: Miami Conservancy
40 N. Main Street, Suite 760
Dayton, OH, 45423
937-223-1271
info@conservancy.com

OKLAHOMA

Kansas Stream Chemistry Database

Description: The Kansas Stream Chemistry Database contains some 750,000 data points describing water quality conditions at more than 500 stream locations in Kansas and bordering regions of adjoining states. The data is stored in DB2 format on an IBM AS-400 computer system maintained by the Kansas Department of Health and Environment.

Database URL: Not available

Chemical groups: Nutrients, other inorganics, radionuclides, general physiochemical, pesticides and SOCs, metals, and particulates.

Period of Performance: 07/05/1967 - 12/31/2001

Contact: Mike Butler or Anthony Stahl
Kansas Department of Health and Environment
1000 SW Jackson
Topeka, KS, 66612-1367
785-296-6603
mbutler@kdhe.state/ks.us or tstahl@kdhe.state.ks.us

Arkansas STORET

Description: This data has been developed over the past 30 years as part of the routine monitoring of the state's waterbodies.

Database URL: Not available

Chemical groups: Nutrients, VOCs, other inorganics, general physiochemical, pesticides and SOCs, general organics, metals, and particulates.

Period of Performance: 01/01/1970 - present

Contact: Alan Price
Arkansas Department of Environmental Quality
8001 National Drive

Little Rock, AR, 72219
501-682-0744
price@adeq.state.ar.us

Oklahoma Environmental Information Management Suite (EIMS)
Description: In 1992 the Groundwater Protection Council (GWPC) began developing the Environmental Information Management Suite (EIMS) to improve regulatory decision-making. This is designed to help states and industry manage natural resources. OK EIMS is one of those resources. Includes USEPA STORET data, private well sampling results from the Oklahoma Water Resources Board (WRB), and Department of Environmental Quality's (DEQ) public water supply cataloging efforts to date.
Database URL: http://virtuales.com/okgis/cfm/login.cfm
Chemical groups: Nutrients
Period of Performance: Not available
Contact: Jon Craig
OK DEQ - Water Quality Division
P.O. Box 1677
Oklahoma City, OK, 73101-1677
405-702-8100
jon.craig@deqmail.state.ok.us

OREGON
Oregon Safe Water System: Chemical Detections, Burns Water Department
Description: Access to data on public water systems in Oregon related to coliform testing, chemical testing, violations, and basic system information.
Database URL: http://159.121.24.16/
Chemical groups: VOCs, other inorganics, radionuclides, pesticides and SOCs.
Period of Performance: Not available
Contact: John Davis
Oregon Department of Human Services Drinking Water Program
PO BOX 14350,
Portland, OR, 97293
503-731-4381
john.a.davis@state.or.us

Laboratory Analytical Storage and Retrieval (LASAR) Database
Description: Provides select parameters by chemical group for data retrieval, interactive mapping of Monitoring Stations, monitoring data by station, event, sub-basin or watershed.
Database URL: http://www.deq.state.or.us/wq/lasar/LasarHome.htm
Chemical groups: Other inorganics, general physiochemical, pesticides and SOCs, general organics, and metals.
Period of Performance: 01/01/1949 - present

Contact: Jeff Jones
Oregon Department of Environmental Quality, Laboratory Division
1712 SW 11th Ave
Portland, OR, 97201-3207
503-229-5983 x257
jones.jeff@deq.state.or.us

Oregon Water Quality Index
Description: Stores index values for each stream monitored. The index is calculated by integrating eight selected water quality parameters (temperature, dissolved, oxygen, biochemical oxygen demand (BOD), pH, ammonia+nitrate nitrogen, total phosphates, total solids, fecal coliform).
Database URL: http://www.deq.state.or.us/lab/WQM/WQI/wqimain.htm
Chemical groups: Nutrients, general physiochemical, and general organics.
Period of Performance: 01/01/1985 - present
Contact: Curtis Cude
Oregon DEQ
Laboratory Division, Water Quality Monitoring Section
811 SW Sixth Avenue
Portland, OR, 97204-1390
503-229-5983/272
labweb@deq.state.or.us

PENNSYLVANIA
Alliance Citizen Monitoring Program Database
Description: Contains basic water quality and nutrients data
Database URL: http://www.acb-online.org/monitoring/site.cfm
Chemical groups: Nutrients and general physiochemical.
Period of Performance: 01/01/1985 - present
Contact: Alliance for the Chesapeake Bay
6600 York Road, Suite 100
Baltimore, MD, 21212
410-377-6270
acbmail@acb-online.org

Susquehanna River Basin Commission (SRBC) Nutrient Assessment Program
Description: Samples are collected at each site during periods of low flow (base flow) and high flow (storm flow) to measure nutrient and suspended sediment concentrations.
Database URL: http://www.srbc.net/nutrientprogram.htm
Chemical groups: Nutrients and general physiochemical.
Period of Performance: Not available
Contact: Susquehanna River Basin Commission

1721 N. Front Street
Harrisburg, PA, 17102
717-238-0423
srbc@srbc.net

Pennsylvania Spatial Data Access (PASDA) The Pennsylvania Geospatial Data Clearinghouse
Description: This clearinghouse provides for sharing of geospatial data, eliminates the creation of redundant data sets, and serves as a resource for locating data throughout the Commonwealth through its data storage, interactive mapping/web GIS applications, and metadata.
Database URL: http://www.pasda.psu.edu/
Chemical groups: Not available
Period of Performance: Not available
Contact: Maurie Caitlin Kelly
PASDA Coordinator
PASDA, The Pennsylvania State University
University Park, PA, 16804
814-863-0104
mck4@psu.edu

Schuylkill and Delaware River Watershed Database
Description: Data was downloaded from the STORET, USGS NASQAN and USGS NAWQA websites, as well as data compiled from local watershed organizations and water suppliers. This data is most robust for basic physical parameters, nutrients, and metals. Water supplier data provides a number of daily or weekly measurements for the past decade. Data spans the period of 1970 to 2002. Updates are made yearly to the database.
Database URL: Not available
Chemical groups: Nutrients, VOCs, other inorganics, radionuclides, general physiochemical, pesticides and SOCs, general organics, metals, and particulates.
Period of Performance: 01/01/1970 - 01/01/2002
Contact: Christopher S. Crockett
Philadelphia Water Department
1101 Market St.
Philadelphia, PA, 19107
215-685-6234
chris.crockett@phila.gov

RHODE ISLAND

Rhode Island Public Water Inventory
Description: Sources of public drinking water, along with their sampling history and location information.
Database URL: Not available
Chemical groups: Nutrients, VOCs, other inorganics, radionuclides, general physiochemical,

pesticides and SOCs, general organics, metals, and particulates.
Period of Performance: 01/01/1970 - 04/29/2002
Contact: Deborah LaFleur
RI Department of Health
3 Capitol Hill, Room 209
Providence, RI, 02908-5097
4012226867
debl@doh.state.ri.us

New England Interstate Water Pollution Control Commission (NEIWPCC)
Description: Contains nutrient and general physiochemical data for lakes, rivers, reservoirs, ponds, and streams.
Database URL: Not available
Chemical groups: Nutrients and general physiochemical.
Period of Performance: 01/01/1990 - 01/01/1999
Contact: Beth Card
New England Interstate Water Pollution Control Commission (NEIWPCC)
Boott Mills South
100 Foot of John Street
Lowell, MA, 01852
978-323-7929
Bcard@neiwpcc.org

SOUTH CAROLINA

South Carolina Ambient Ground Water Quality Monitoring Network
Description: Compilation of water quality data collected from 115 wells comprising nine major aquifers of the state.
Database URL: http://www.scdhec.net/eqc/water/pubs/amb2000.pdf
Chemical groups: Nutrients, other inorganics, general physiochemical, general organics, metals, and particulates.
Period of Performance: 01/01/1987 - present
Contact: Chad Reihm
Bureau of Water, South Carolina Department of Health and Environmental Control
2600 Bull Street
Columbia, SC, 29201
803-898-3799
reihmcr@dhec.state.sc.us

South Carolina Groundwater Contamination Inventory
Description: Inventory of groundwater chemical data including metals, VOCs, radionuclides, and organics.
Database URL: http://www.scdhec.net/eqc/water/html/gwci01/2001gwci.html

Chemical groups: Nutrients, VOCs, radionuclides, pesticides and SOCs, and metals.
Period of Performance: 01/01/1980 - present
Contact: Harriet Gilkerson
Bureau of Water, South Carolina Department of Health and Environmental Control
2600 Bull Street
Columbia, SC, 29201
803-898-3796
gilkerhh@dhec.state.sc.us

TEXAS
Source Water Assessment Program (SWAP) Potential Source of Contamination (PSOC) Data set
Description: Data set (composite of 40+ statewide data sets of potential contaminants) consists of point locations for potential contaminant sources of various regulated chemical constituents monitored by public drinking water systems.
Database URL: Not available
Chemical groups: Nutrients, VOCs, other inorganics, radionuclides, general physiochemical, Pesticides and SOCs, general organics, and metals.
Period of Performance: 10/01/1986 - 05/31/2002
Contact: Sean Ables
Texas Natural Resources Conservation Commission
P.O. Box 13087
Austin, TX, 78711
512-239-1758
sables@tnrcc.state.tx.us

Arkansas STORET
Description: This data has been developed over the past 30 years as part of the routine monitoring of the state's waterbodies.
Database URL: Not available
Chemical groups: Nutrients, VOCs, other inorganics, general physiochemical, pesticides and SOCs, general organics, metals, and particulates.
Period of Performance: 01/01/1970 - present
Contact: Alan Price
Arkansas Department of Environmental Quality
8001 National Drive
Little Rock, AR, 72219
501-682-0744
price@adeq.state.ar.us

Texas Water Development Board (TWDB) Groundwater Database (GWDB)
Description: Access to water-level measurements and groundwater quality analyses for over

123,000 water wells, springs and monitoring wells.
Database URL: http://www.twdb.state.tx.us/data/waterwells/well_info.html
Chemical groups: Nutrients, general physiochemical, and metals.
Period of Performance: 01/01/1960 - present
Contact: Roger Quincy
Texas Water Development Board
P.O. Box 13231
Capitol Station
1700 N. Congress Avenue
Austin, TX, 78711-3231
512-936-0833
roger.quincy@twdb.state.tx.us

Texas Natural Resources Conservation Commission (TNRCC) – Sampling Data Query, Surface Water Quality Monitoring
Description: The event data file provides station information, sample date and time, sample depth, and the entity that performed the sampling. The result data file provides the STORET and measurement information.
Database URL: http://www.twdb.state.tx.us/data/waterwells/well_info.html
Chemical groups: Not available
Period of Performance: 01/01/1993 - present
Contact: Texas Natural Resource Conservation Commission (TNRCC)
PO Box 13087
Austin, TX, 78711-3087
512-239-DATA
wquality@tnrcc.state.tx.us

UTAH
Underground Storage Tanks (UST), Leaking Underground Storage Tanks (LUST), Brownfields, Superfund, Toxics Release Inventory (TRI), Voluntary Clean Program (VCP)
Description: Click on Interactive Map link. Contains data about sites regulated by the Division of Environmental Response.
Database URL: http://www.deq.state.ut.us/EQERR/index.htm#
Chemical groups: Not available
Period of Performance: 01/01/2002 - 02/19/2002
Contact: Harold Sandbeck
Utah Department of Environmental Quality (UDEQ) / DERR
PO Box 144840
Salt Lake City, UT, 84114-4840
801-536-4100

Statewide Geographic Information Database
Description: Numerous collections of data regarding land use in Utah.
Database URL: http://agrc.utah.gov/sgid/statewide/index.htm
Chemical groups: Not available
Period of Performance: 01/01/1990 - 02/20/2002
Contact: Dennis Goreham
Automated Geographic Reference Center
5130 State Office Building
Salt Lake City, UT, 84114
801-538-3163
dgoreham@gis.state.ut.us

VERMONT
Water Supply Division (WSD) Inventory
Description: Chemical and bacteriological monitoring data required under the federal Safe Drinking Water Act.
Database URL: Not available
Chemical groups: Nutrients, VOCs, other inorganics, radionuclides, general physiochemical, pesticides and SOCs, metals, and particulates.
Period of Performance: 01/01/1989 - 03/13/2002
Contact: Jean Nicolai
Agency of Natural Resources, Department of Environmental Conservation
103 South Main Street
Waterbury, VT, 05671-0403
802 241 3405
jeann@dec.anr.state.vt.us

New England Interstate Water Pollution Control Commission (NEIWPCC)
Description: Contains nutrient and general physiochemical data for lakes, rivers, reservoirs, ponds, and streams.
Database URL: Not available
Chemical groups: Nutrients and general physiochemical.
Period of Performance: 01/01/1990 - 01/01/1999
Contact: Beth Card
New England Interstate Water Pollution Control Commission (NEIWPCC)
Boott Mills South
100 Foot of John Street
Lowell, MA, 01852
978-323-7929
Bcard@neiwpcc.org

VIRGINIA

Chesapeake Bay Program Water Quality Database

Description: The database includes measured and calculated physical, metal, other organics, general inorganics and particulates for 49 mainbay and approximately 150 tributary stations for Maryland, Virginia, and the District of Columbia.

Database URL: http://cobia.chesapeakebay.net/data/data_desc.cfm?DB=CBP_WQDB

Chemical groups: Nutrients, other inorganics, general physiochemical, general organics, and metals.

Period of Performance: 01/01/1984 - present

Contact: Ricky Bahner
Chesapeake Bay Program,
410 Severn Ave., Suite 109
Annapolis, MD, 21403
800-9687-229 ext. 742
rbahner@chesapeakebay.net

Chesapeake Bay Toxics Program Database

Description: This database stores pesticides, SOCs, VOCs, metals, other inorganics, and general organics data.

Database URL: http://cobia.chesapeakebay.net/data/data_desc.cfm?DB=CBP_TOXDB

Chemical groups: VOCs, other inorganics, pesticides and SOCs, general organics, and metals.

Period of Performance: 01/01/1973 - present

Contact: Kelly Shenk
Chesapeake Bay Program
410 Severn Ave., Suite 109
Annapolis, MD, 21403
800-9687-229 ext. 728
shenk.kelly@epa.gov

Alliance Citizen Monitoring Program Database

Description: Contains basic water quality and nutrients data.

Database URL: http://www.acb-online.org/monitoring/site.cfm

Chemical groups: Nutrients and general physiochemical.

Period of Performance: 01/01/1985 - present

Contact: Alliance for the Chesapeake Bay
6600 York Road, Suite 100
Baltimore, MD, 21212
410-377-6270
acbmail@acb-online.org

Occurrence and Distribution of Pesticides in Chesapeake Bay
Description: Compiles pesticide data measured in surface water, surface microlayer, groundwater, sediment, and biota of the Bay. Twelve pesticides are included in the database: alachlor, atrazine, carbofuran, chlordane, chlorpyrifos, chlorothanonil, diflubenzuron, linuron, metolachlor, permethrin, simzine, and toxaphene. Summary statistics with number of samples collected, maximum, minimum, and mean is provided for each station.
Database URL: http://www.agnic.org/cbp/index.html
Chemical groups: Pesticides and SOCs.
Period of Performance: 01/01/1976 - 01/01/1994
Contact: W. Edward Johnson
National Oceanic and Atmospheric Administration (NOAA)
1325 East West Hwy 8176 SSMC2
Silver Spring, MD, 20910-3283
301-713-0177
Ed.Johnson@noaa.gov

WASHINGTON
Environmental Assessment Program – River Stream and Water Quality Monitoring
Description: Access to samples from about 82 stations per year. About 20 "basin" stations are monitored for one year (sometimes returning every five years) and 62 "long-term" stations are monitored every year. This also organizes data by Watershed Resource Inventory Areas (WRIAs) which refer to the state's major watershed basins.
Database URL: http://www.ecy.wa.gov/programs/eap/fw_riv/rv_main.html
Chemical groups: Nutrients, general physiochemical, and particulates.
Period of Performance: 01/01/1959 - 01/01/2001
Contact: Dave Hallock
Washington State Department of Ecology, Environment Assessment
PO Box 47600
Olympia, WA, 98504-7600
360-407-6681
daha461@ecy.wa.gov

Pesticides in Public Supply Wells of Washington State
Description: Pesticide data collected from wells in the Puget Sound Basin and in the Central Columbia Plateau.
Database URL: http://wa.water.usgs.gov/ccpt/pubs/fs-122-96.html
Chemical groups: Nutrients, pesticides and SOCs.
Period of Performance: 01/01/1995 - 01/01/1998
Contact: Project Chief
Central Columbia Plateau NAWQA Study
1201 Pacific Avenue, Suite 600
Tacoma, WA, 98402

206-593-6530/235
nawqa_ccpt_wa@usgs.gov

WEST VIRGINIA

West Virginia Source Water Assessment and Wellhead Protection Program SWAP data online*

Description: The West Virginia Source Water Assessment and Wellhead Protection Program will provide access to online data concerning all of the surface water intakes for public water in the state of West Virginia. This will also include background materials, source water assessment materials, fact sheets, documents, and maps.
Database URL: Not available
Chemical groups: Not available
Period of Performance: Not available
Contact: Bill Toomey
West Virginia Department of Health
815 Quarrier Street, Suite 418
Charleston, WV, 25301
304-558-2981
wtoomey@wvdhhr.org
This data set is currently under development

WISCONSIN

Drinking Water System

Description: Includes all SDWA and other drinking water monitoring results.
Database URL: http://www.dnr.state.wi.us/org/water/dwg/DWS.htm
Chemical groups: Nutrients, VOCs, other inorganics, radionuclides, general physiochemical, pesticides and SOCs, general organics, metals, and particulates.
Period of Performance: 05/03/1992 - 05/03/2002
Contact: Don Swailes
Wisconsin Department of Natural Resources
101 S. Webster Street
Madison, WI, 53703
(608) 266-7093
swaild@dnr.state.wi.us

WYOMING

Statewide Water Resources Data Inventory

Description: It is a water resources data inventory including several data sets such as STORET, NPDES, State Soil Geographic (STATSGO), National Hydrographic Data set (NHD), Permit Compliance System (PCS), and others.
Database URL: http://waterplan.state.wy.us/sdi/sdi.html
Chemical groups: Nutrients, VOCs, other inorganics, radionuclides, general Physiochemical,

pesticides and SOCs, general organics, metals, and particulates.
Period of Performance: Not available
Contact: Wyoming Water Development Commission
Herschler Building, Fourth Floor West
Cheyenne, WY 82002
307-777-7626
waterplan@waterplan.state.wy.us

REFERENCES

United States Environmental Protection Agency (USEPA), Office of Water, 1997. State Source Water Assessment and Protection Programs Final Guidance, 816-R-97-009, August, 1997.

ABBREVIATIONS

AWWA	American Water Works Association
AwwaRF	Awwa Research Foundation
AWWSC	American Water Works Service Company
BASINS	Better Assessment Science Integrating Point and Nonpoint Sources
BOD	Biological Oxygen Demand
CHL-A	Chlorophyll *a*
CWA	Clean Water Act
DBP	Disinfection By-Product
DEP	Department of Environmental Protection
DEQ	Department of Environmental Quality
DO	Dissolved Oxygen
DOC	Dissolved Organic Carbon
DNR	Department of Natural Resources
DWRP	Drinking Water and Radiological Protection Division
EBMUD	East Bay Municipal Utility District
EIMS	Environmental Information Management Suite
EMPACT	Environmental Monitoring for Public Access and Community Tracking
FOIA	Freedom of Information Act
FRDS	Federal Reporting Data System
GIS	Geographic Information System
GWMAP	Groundwater Monitoring and Assessment Program
GWIS3	Generalized Water Information System
GWPC	Groundwater Protection Council
GWDB	Groundwater Database
HUC	Hydrologic Unit Code
ICR	Information Collection Rule
IOC	Inorganic Chemical
LASAR	Laboratory Analytical Storage and Retrieval
LDC	Legacy Data Center

LT2	Long-Term 2 Enhanced Surface Water Treatment Rule
LUST	Leaking Underground Storage Tanks
MBTE	Methyl Tert-Butyl Ether
MCL	Maximum Contaminant Level
MDE	Maryland Department of the Environment
MDH	Minnesota Department of Health
MDEQ	Michigan Department of Environmental Quality
NASQAN	National Stream Quality Accounting Network
NAWQA	National Water Quality Assessment Program
NCOD	National Contaminant Occurrence Database
NEIWPCC	New England Interstate Water Pollution Control Commission
NHD	National Hydrographic Data set
NOAA	National Oceanic and Atmospheric Administration
NPDES	National Pollutant Discharge Elimination System
NRIS	Natural Resource Information System
NWIS	National Water Information System
PAC	Project Advisory Committee
PASDA	Pennsylvania Spatial Data Access
PCB	Polychlorinated Biphenyl
PCE	Tetrachloroethylene
PCS	Permit Compliance System
PEARL	Public Educational Access to Environmental Information in Maine
PSOC	Potential Source of Contamination
PWD	Philadelphia Water Department
PWSS	Public Water System Supervision
QA/QC	Quality Assurance/Quality Control
SDWA	Safe Drinking Water Act
SDWIS	Safe Drinking Water Information System
SMCL	Secondary Maximum Contaminant Level
SOC	Synthetic Organic Compound
SRBC	Susquehanna River Basin Commission
SRF	State Revolving Fund
STATSGO	State Soil Geographic
STORET	Storage and Retrieval
SWAP	Source Water Assessment Program

TCE	Trichloroethylene
TDS	Total Dissolved Solids
THMs	Trihalomethanes
TMDL	Total Maximum Daily Load
TNRCC	Texas Natural Resources Conservation Commission
TOC	Total Organic Carbon
TPMC	Technology Planning and Management Corporation
TSS	Total Suspended Solids
TRI	Toxics Release Inventory
TWDB	Texas Water Development Board
UDEQ	Utah Department of Environmental Quality
URL	Uniform Resource Locator
USEPA	United States Environmental Protection Agency
USGS	United States Geological Survey
UST	Underground Storage Tank
VCP	Voluntary Cleanup Program
VOC	Volatile Organic Compound
WATERS	Watershed Assessment, Tracking & Environmental Results
WRB	Water Resources Board
WRIA	Watershed Resource Inventory Areas
WSD	Water Supply Division

IWA Publishing's authorised EU representative for General Product Safety Regulations is Diane D'Arras, 15 rue Duret, 75116 Paris, France, e-mail: safety@iwap.co.uk.

Printed and bound by CPI Group (UK) Ltd, Croydon, CR0 4YY
17/09/2026
02218808-0012